F. R. Siegel

Environmental Geochemistry of Potentially Toxic Metals

Springer

Berlin
Heidelberg
New York
Barcelona
Hong Kong
London
Milan
Paris
Tokyo

Frederic R. Siegel

Environmental Geochemistry of Potentially Toxic Metals

With 40 Figures and 33 Tables

 Springer

Professor Dr. Frederic R. Siegel
The George Washington University
Department of Earth
and Environmental Sciences
Washington D.C. 20052
USA

ISBN 3-540-42030-4 Springer-Verlag Berlin Heidelberg New York

Cip Data applied for
Die Deutsche Bibliothek – CIP-Einheitsaufnahme

Siegel, Frederic R.:
Environmental geochemistry of potentially toxic metals : with 33 tables /
Frederic R. Siegel. – Berlin ; Heidelberg ; New York ; Barcelona ; Hongkong ;
London ; Mailand ; Paris ; Tokio : Springer, 2002
   ISBN 3-540-42030-4

Springer-Verlag is a company in the BertelsmannSpringer publishing group.

http://www.springer.de

© Springer-Verlag Berlin Heidelberg 2002
Printed in Germany

Cover design: E. Kirchner, Heidelberg
Typesetting: Fotosatz-Service Köhler GmbH, Würzburg
Printed on acid-free paper   SPIN: 10796556   32/3130/as 5 4 3 2 1 0

Preface

Contamination of the earth's ecosystems by potentially toxic metals/metalloids is a global problem. It will likely grow with our planet's increasing populations and their requirements for natural resources (e.g., water, food, energy, waste-disposal sites) and metals-based goods. The health impacts of pollution from the ingestion of heavy metals/metalloids via respiration, food, and drinking water are most often long-term and manifest themselves in many ways. These include, for example, disminution of mental acuity, loss of motor control, critical organ dysfunction, cancer, chronic illnesses and concomitant suffering, incapacitation, and finally death. The incidence and geographic distribution of disease (epidemiology) has been well-documented historically and in modern times for toxic metals-triggered diseases in humans, animals and vegetation.

The role of the environmental geochemist and colleagues in environmental sciences is to scientifically evaluate how to manage metals/metalloids at sources or *in-situ* so as to alleviate or eliminate their negative health impacts on living populations. This is initiated by identifying sources and by developing models of the physical, chemical and biological controls on mobilization, interaction, deposition and accumulation of potentially toxic metals/metalloids in source systems and earth ecosystems. From this knowledge base, environmental scientists (e.g., geologists, chemists, biologists, environmental engineers, physicists/meteorologists) work together to develop

concepts and technological methodologies to preserve global ecosystems. Their concerted efforts are equally focussed on devising
strategems to remediate ecosystems still carrying heavy metals/metal-
loids pollutant burdens from ancient and modern societies.

This book brings an appreciation of the complexity involved in
studies on potentially toxic metals to scientists who investigate
chemical/biochemical metals pollution problems in the earth's vast
array of living environments. This is done in initial chapters that
focus on heavy metals/metalloids and their roles as essential elements
and as pollutants, identify metals' sources, characterize metals'
mobility/immobility in environmental media, and establish their
pathways, geochemical cycles and bioaccumulation in ecosystems.
Subsequent chapters deal with defining contamination values and
processes that can affect them through time, and with assessing ecosystems' health statuses via various indicator media and their chemical analysis under proper protocols. The final chapters of the
book discuss remediation/alleviation strategies and environmentally-
reasoned decision-making to keep earth systems sustainable.

Frederic R. Siegel
Washington, D.C.
April, 2001

Dedication

To dedicate this book is to acknowledge many sources of support that kept me on an even keel when writing it, provided me with insight as to what should be emphasized in it, and allowed me access to knowledge garnered during research projects worldwide. Thus, I dedicate this book first to my wife, Felisa, my daughters, Gabriela (spouse Morris Benveniste) and Galia, and my grandchildren, Naomi, Coby and Noa. I dedicate this book also to the myriad of undergraduate and graduate students from many parts of the world I have had in classes and with whom I worked on M.S. thesis and Ph.D. dissertation research projects. Their questions and discussions in and out of the classroom were important in formulating what should go into the manuscript. Finally, I dedicate this book to the broad spectrum of scientists working on environmental problems whose published research expanded my own fields of knowledge and contributed to the breadth of the book.

Table of Contents

CHAPTER 1

Geochemistry in Ecosystem Analysis of Heavy Metal Pollution

Environmental Geochemistry

Environmental geochemistry is the discipline that uses the chemistry of the solid earth, its aqueous and gaseous components, and life forms to assess heavy metal contamination impacts on our planet's ecosystems. It deals with physical, chemical and biological conditions in an environment such as temperature, state of matter, acidity (pH), reduction-oxidation (redox) potential, bacterial activity and biological oxygen demand (BOD). These factors and others influence the mobilization, dispersion, deposition, distribution and concentration of potentially toxic metals/metalloids which can impair the health of organisms in an ecosystem. As such, environmental geochemical data identify pristine chemical conditions that pose no threats to ecosystem inhabitants, those that may suffer from (natural) chemical intrusion from rock weathering and decomposition, and environments that are at risk from chemical element pollution as a result of human activities.

Often the chemical damage to an environment originates from a combination of natural and anthropogenic input. The physical and biological processes observed and measured in "at risk" environments complement the geochemical data used to assess the risk posed to an ecosystem mainly by higher than natural baseline contents of potentially toxic metals. The same data can be useful to locate possible origin(s) of the metals. Although most emphasis is on pollution threats from high metal concentrations, very low contents of metals that are essential micronutrients can cause deficiency toxification in organisms and should not be overlooked in contaminant studies.

At its best, environmental geochemistry first predicts areas that could be at risk from natural (rock chemistry) and anthropogenic (planned development projects) chemical intrusions. Beyond defining and identifying chemical threats to an ecosystem, environmental geochemistry extends into the realm of assessing physical, chemical and biological remediation technologies to extract, isolate, contain and dispose of residues of past chemical intrusions. In this way, the input of metals in concentrations that can harm an environment and its inhabitants can be eliminated or at least greatly alleviated at their source(s).

Potentially Toxic Metals in Focus

The array of chemicals analysed in environmental projects includes radioactive elements, organic and organo-metallic compounds, and heavy metals and metalloids and their chemical species. In this text, the assessment of environments is limited to heavy metals and metalloids categorized as potentially toxic to life forms by organizations such as the U.S. Environmental Protection Agency, the World Health Organization and the Arctic Monitoring and Assessment Programme. The metals/metalloids considered here are As, Be, Cd, Co, Cr, Cu, Fe, Hg, Mn, Mo, Ni, Pb, Sb, Sc, Se, Ti, Tl, V and Zn. Other elements that are major components in rock-forming minerals such as Al, Ca, K, Mg, Na, P and Sr are considered to some degree since they can help identify a solid phase that may contain a potential pollutant. With additional

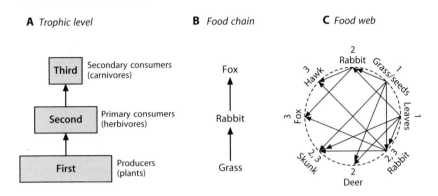

Fig. 1-1. The transfer of nutrients (and energy) is represented by A) trophic levels, B) a food chain, and C) a food web. The numbers in c) refer to the tropic levels occupied by each member of the food web (after Mackenzie, 1998)

information, likely modes of transport and pathways to depositional environments may be established. The main modes of contaminant access to organisms is through food, water and the atmosphere. The terms food chain, food web and trophic levels are used to describe pollutant pathways to organisms. The food chain can be defined as an ordering of the organisms in an ecosystem following a predation sequence in which each has the next lower member as a food source. A food web describes the totality of interacting food chains in an ecosystem. A trophic level is a hierarchical strata of a food web characterized by organisms which are the same number of steps removed from the primary producers. Figure 1-1 illustrates these relations. The interactions of biota along a food web in a natural environment are shown in Figure 1-2 by the composite food web for the Arctic marine ecosystem. Ingested heavy metals move up food chains and through food webs to higher trophic levels and may bioaccumulate and be biomagnified along this pathway. There is a threat to the health of feeders at higher trophic levels from the ingestion and bioaccumulation over time of some metals that reach critical concentrations toxic to body fluids, tissue, or specific organs.

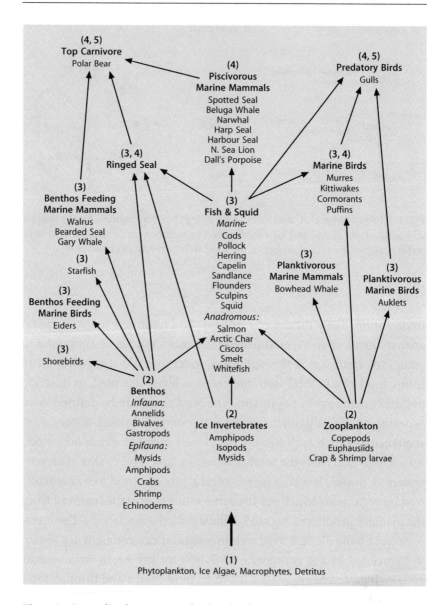

Fig. 1-2. Generalized composite food web of the Arctic marine ecosystem. The parenthesized numbers represent trophic level in ascending order with examples of each major category of biota (after Becker, 1993)

Important Factors That Affect Environmental Impact

There are many documented cases in which long-term ingestion of bioavailable heavy metals (e.g., Hg, Cd, As, Pb) in an ecosystem have built up toxic metal concentrations in organisms and caused disease and death. In other instances, the lack of an essential micronutrient heavy metal (e.g., Se, Zn, Cu) in an organism's food supply has also caused illness and death. The controls on metal input to an ecosystem and output from it, balanced or not, is inherently a function of the geological environment and processes active there mediated by physical, chemical and biological factors. The chemistry of a natural geosystem may be strongly affected by heavy metals loading in a living environment that result from human activities. Awareness of this latter facet of environmental intrusion allows pre-planning during the evaluation of an economic development project (e.g., chemical, mining, electricity generation). The inclusion of effluent and emissions control systems and carefully designed waste disposal sites will avoid a future damaging toxic metals input.

Various factors affect the mobilization, dispersion, deposition and concentration of elements from rock-forming and ore minerals and hence their effects on the livability of a natural environment. In addition to chemical parameters such as element solubility, concentration and speciation, these include pH, redox potential, temperature, BOD, salinity, particulate size, mineralogy, sorption, and bioaccumulation as most important. A vivid example of the effect of changes in one of these factors, pH, on life in a riverine environment is illustrated in Figure 1-3. This figure illustrates that at "normal" fluvial water conditions of pH = 6–7, the ecosystem supports the 15 species represented (fish, amphibians, shellfish, insect). However, as the pH falls to more intense acidic conditions from acid mine drainage, acid rock drainage and/or acid rain and input is maintained over time, the number of species that survive falls dramatically until pH = 4. At this point, the riverine system supports none of the species but may support bacteria

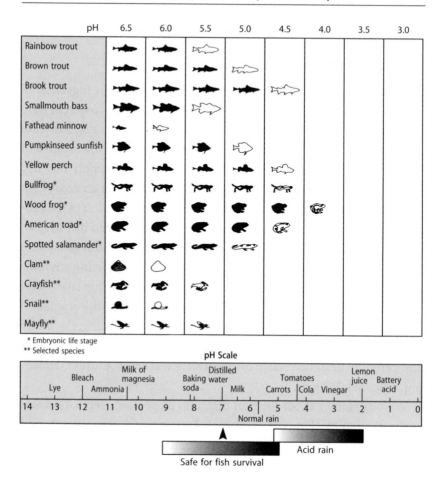

* Embryonic life stage
** Selected species

Fig. 1-3. The effect of pH change on survivability of some organisms in a riverine regime. The pH scale indicates the range of water pH considered safe for fish to survive and the pH range of acid rain for comparison. The approximate pH at which an organism dies with extended exposure is given by the organism pattern in white (after Christensen, 1991, from U.S. Department of Interior, Fish and Wildlife Service)

and algae. With strongly acidic conditions, many heavy metals are mobilized and dispersed downstream where some may be immobilized by adsorption onto Fe and Mn oxy/hydroxides and/or absorbed and concentrated by algae or other forms at the lower trophic level. These metals may bioaccumulate in the food web to endanger life forms which biomagnify them to critical levels. Organisms feeding on metal-laden "foods" may develop physical and bodily disorder symptoms or conditions indicative of pollution.

Sudden events may temporally extinguish a riverine ecosystem as a result of the intrusion of a non-metal component. In some cases potentially toxic metals may be released simultaneously with the non-metal into streams and rivers. Depending on the dilution with dispersion of the metals and subsequent deposition, the metals may impair the livability of an ecosystem for an extended period of time beyond that of the non-metal pollutant. An example of such a scenario took place on January 30, 2000. There was a sudden overflow of CN^- bearing waters from a dam in the mining center of Baia Mare, Romania where the CN^- heap leach technique was being used to extract Au from old mine tailings and residues. The CN^- contaminated water together with a heavy metals load, flowed into the Tiza river killing all life forms, likely by oxygen deprivation caused by the CN^-. More than 100 tons of dead and bloated fish and other dead aquatic life (including bacteria) were removed from the river. Birds and small mammals fell to the CN^- and heavy metals-laced waters as well. From Romania, the Tiza river flows into Hungary and Yugoslavia. About 400 km of the river were affected. The degradation of CN^- was slow because of low temperatures and an ice cover that limited its exposure to sunlight. At Zemun, just northwest of Belgrade there is a confluence with the Danube river. The World Health Organization (WHO) maximum allowable concentration of CN^- in mine waste water is 0.07 ppm but the outflow from the heap leach pond measured 1.1 ppm and reached as high as 2 ppm. By February 12, CN^- levels just north of Belgrade were below 0.2 ppm. The mass and state of heavy metals released into the river has to be determined because of possible inflow to aquifers as well as concentrations in sediments and bioavailability from them. The metal assemblage, likely with high contents of Cd and Pb, may create or compound pollutant problems in the riverine environments for some time after the CN^- is no longer a cause for concern, depending on pH, redox potential and other ambient conditions.

The interdependency of physical factors becomes apparent when the temperature of an environment is considered. Warm waters hold less gas than cold waters. Thus, "warm" water fluvial systems and lakes have less capacity to hold CO_2 gas and have a higher pH (less acidic) than cold waters which will have a more acidic pH under otherwise the same environmental conditions. Similarly, warmer waters will hold less oxygen. Input of biodegradable matter that has less of a BOD than an aqueous system, will not diminish a system capacity to sustain life. However, if an input of organic matter is greater than the BOD capacity of an aqueous system to decompose it, oxygen-depleted conditions can develop. Over time this can lead to eutrophication and the reduced capability of a water body to support an original array of life forms.

Essentiality of and Risks from Potentially Toxic Metals

There are several heavy metals that are micro- or macro-nutrients essential to maintain a good health status in humans and other organisms. Table 1-1 categorizes elements that must be ingested regularly for the various human body organs and systems to function optimally. In some cases, organisms need a particular chemical species of an element for biological systems to operate normally. Synergistic and antagonistic relations between metals/metalloids can also influence the impact of heavy metal nutrients in a living ecosystem. In humans, for example, toxic effects from ingestion and bioaccumulation of As may be damped by the antagonist element Se. However, potentially toxic metal interactions can cause problems for humans as well. For

Table 1-1. Essential micronutrients for optimal functioning of biological processes and organs in humans and non-essential heavy metals (compiled from Merian, 1991; Fergusson, 1990; Mertz, 1981). Data for essential macronutrients and other micronutrients are given for comparison (from Crounse et al., 1983)

Essential heavy metal micronutrients (a few mg or µg per day):	As, Co, Cr, Cu, Fe, Mn, Mo, Se, V, Zn
Non-essential heavy metals:	Be, Cd, Hg, Ni, Pb, Sb, Sn, Ti
Macronutrients (~ 100 mg or more per day):	Ca, Cl, Mg, P, K, Na, S
Other essential micronutrients:	F, I, Si

Ni and Sn may be essential micronutrients (Crounse et al., 1983).

example, ingestion of Pb can interfer with the absorption of Fe in the intestine thereby causing an Fe deficiency. The Pb can also increase Cu and Ca nutrient deficiencies that already exist (Fergusson, 1990).

There are ranges of concentrations of bioavailable, yet potentially toxic, essential metals/metalloids that are needed in micro-quantities by critical organs and for biochemical processes. When there is an excess or a deficiency in the diet of one or more than one of these elements over time, an organism can develop an abnormal condition, disease, or even die. For example, a deficiency of Se in humans, especially young children, may be the cause of congestive heart failure (Keshan disease). On the other hand, ingestion of excess amounts of Se can result in acute vascular disruption and hemorrhaging and/or chronic dermatitis, hair loss, jaundice and caries (Fergusson, 1990). Figure 1-4 A gives a general dose/response graph that is self-explanatory in illustrating the relation between the concentration of an essential element ingested and the health status of an organism.

Other potentially toxic metals are non-essential (Table 1-1) and cause no problems if missing from a diet. Likewise, ingestion of these metals may have no effects on an organism if they are bioevacuated without any bioaccumulation or interference with biochemical and

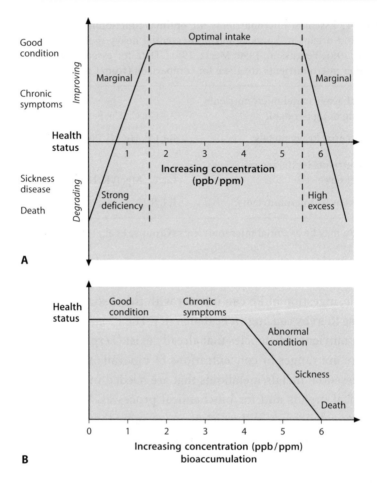

Fig. 1-4. The health response of an organism to: A. Ingested concentration over time of essential micronutrient potentially toxic metals that are required for metabolism and removal of compounds from cells. B. Ingested concentration over time of non-essential potentially toxic metals that bioaccumulate in an organism

organ functions. Conversely, bioaccumulation of one or more potentially toxic non-essential metals can degrade the health status of an organism gradually over the long term with a development of chronic illness and disease symptoms. The degradation may be rapid if large doses are ingested. Non-essential metals/metalloids and their chemical species that cause health problems accumulate in different organs. Methylmercury (CH_3Hg^+), for example, tends to accumulate in the brain whereas Cd^+ and Hg^{2+} have more of an affinity for the kidneys. Figure 1-4B illustrates a dose-response graph for a non-essential potentially toxic metal.

There are threshold concentrations of metals in critical organs or biochemical systems of organisms that cause health of problems. In humans, for example, concentrations >10 µg/g Hg in the liver or >6 µg/g Hg in the brain can cause death whereas >20 µg/dl Hg in the blood can cause chromosome damage (Fergusson, 1990). A body burden of 55 mg Hg for a 51 kg person can cause the onset of loss of motor control (ataxia). The build-up of a metal in an organism can be from ingestion of increases in available amounts in a food web or from bioaccumulation of low contents over a period of time. The result is the same: chronic illness, incapacitation or death. Table 1-2 reviews the essential/non-essential roles of the focus metals.

Epidemiology

Determination of a possible relation between heavy metals/metalloids and disease or death in a human population is made by epidemiologists. They relate the incidence and geographic distribution of diseases to probable causal factors such as the accumulation of excess amounts of metals or the absence of necessary metals in an organism. Chemical analyses are made on a variety of media such as body fluids or excretions (blood, milk, prespiration, saliva) including elimination

Table 1-2. Toxic effects from ingestion of excess or deficiency amounts over time of essential micronutrient or non-essential potentially toxic metals in living populations. The effects may be chronic or acute and include mutagenic (induction of gene mutations), carcinogenic (induction of cancer), teratogenic (developmental defects) in nature and affects metabolism and elimination processes. Compiled from Fergusson (1990) and Merian (1991)

As: Essential. Carcinogen?

Be: Non-essential. Toxic. By inhalation of dust of metal or compounds

Cd: Non-essential. Toxic in soluble and respirable forms; interfers with Zn in enzyme catalysis and key metabolic processes, and Zn bioavailability. Bioaccumulated at all toxic levels. Toxic to some plants at concentrations much lower than Zn, Pb and Cu. Is a carcinogen and is also teratogenic and embryoidal (Smith, 1999).

Co: Essential. Located in active site of cobaltamine (vitamin B_{12}). Plays an important role in biochemical reactions essential for life.

Cr: Essential. Hexavalent via anthropogenic activities toxic. Essential at low concentrations but toxic at elevated levels. Cr^{VI} very soluble, toxic and a carcinogen. 50 µg/l in drinking water. Cr^{III} sparingly soluble and relatively non-toxic. Ilton (1999).

Cu: Essential. Necessary nutrient but excesses can produce toxicity.

Fe: Essential element. Can exert a strong influence on biological reactions.

Hg: Non-essential. Poisoning over time gives neuropathy.

Mn: Essential.

Mo: Essential. Molybdenosis in cattle.

Ni: Essential to plants, possibly essential to animals and humans.

Pb: Non-essential and toxic.

Sb: Non-essential.

Se: Essential at trace concentrations; toxic at elevated concentrations. Selenate more toxic to plants than selenite (Ilton, 1999).

Sn: Non-essential.

Ti: Non-essential.

V: Essential. Beneficial biochemically for protection from tooth decay; may be partly responsible for lung disease (Snyder, 1999).

Zn: Essential trace element. For growth, development and reproduction

products (urine, feces), growing parts (hair, nails, muscles, cartiledge), shedding parts (teeth, skin, bone), and critical organs (kidney, liver, heart, brain, lung, pancreas). In these analyses, representative sample selection, protocol-followed sampling for analysis, and the analytical system used (detection limit, interferences, quality control) are critical to the obtention of reliable data to allow the best interpretion of results.

In some cases, if a cause-effect relation is established between a disease and the ingestion of excess or deficiency concentrations of a potentially toxic metal/metalloid, a therapy may be devised. In China, for example, the link between deficiency of Se in the diet and Keshan disease (cited previously) was proposed (Chen et al., 1980). A weekly supplement of 0.5–1 mg Se greatly reduced the incidence of the disease (Yin et al., 1983). This was a signal event. It alerted medical scientists to the possibility that where there was a high incidence of Keshan disease in other parts of the world, the cause could be a dietary deficiency of Se. The deficiency could be related to the lack of Se in the food web growth environment. Fordyce et al., (2000) studied Se responsive diseases in China and concluded that geology or rock type from which soils formed is the main control on Se and other trace element distributions. They proposed that adsorption by organic matter and complexing with Fe oxyhydroxides restrict Se bioavailability. Johnson et al., (2000) also found that a high organic content of soils can chelate and not release Se to food crops or waters. For some diseases where there seems to be a link to heavy metals, experimental therapy has not been successful as it was in the case for Se and Keshan disease and health damage is irreversible. Therapies to cure or ameliorate health problems thought to originate from excess/deficiency ingestion of heavy metals are being researched. If signal events such as the Se case described above are not communicated to public health officials or are not heeded, chronic illness, abnormal conditions, disease and death which can be avoided will continue to affict some societies.

Role of the Environmental Geochemist

The role of geochemists investigating high contents of potentially toxic metals in an ecosystem begins with an understanding of the probable sources and origins of the metals. The geochemist applies principles of chemical mobility in solid, liquid and gaseous phases to reveal the pathways through which the metals move into ecosystems and their fate once in a living environment. He/she determines natural (baseline) concentrations for each sample type being analysed to monitor the health status of organisms and environments which support them. It is against baseline concentrations that the geochemist can establish the existence of contamination and calculate an enrichment factor or degree of contamination. A knowledge of the response of pollutant metals to imposed physical, chemical and biological conditions is used by geochemists to determine the most efficient, economical, and least socially and culturally disturbing methods of remediating existing toxic metal(s) pollution. This knowledge serves also as a basis for planning new development projects with the purpose of preventing future damaging releases of potentially toxic metal pollutants into an environment.

CHAPTER 2

Sources and Origins of the Metals

Natural Igneous Rocks

The natural sources of metals in the environment lie with the rocks and processes by which they formed and which affected them after lithification. The primary rocks are called igneous and have a wide range of mineral and chemical composition. Minerals are inorganic and each of the more than 3000 minerals known is *unique* in its chemical composition and its orderly internal crystalline structure. Igneous rock minerals form at different cooling stages during the crystallization of a magma (molten rock material within the earth) containing all chemical elements. The union of different minerals or mineral groups and the proportions in which they are present at a given stage of crystallization, together with crystal size (texture) are the basis for classifying the principal igneous rocks. Igneous rocks that form (lithify) as magma cools within the earth slowly are comprised of minerals that solidify into large crystals that can be seen with the naked eye. The rocks are subsequently exposed at the earth's surface by erosion after being uplifted during mountain-forming processes. Igneous rocks that cool rapidly when magma is extruded onto the earth's surface by volcanic activity (e.g., lava) have very fine crystals that can be seen only when such rock is cut into "thin sections" and viewed through a high-power petrographic microscope.

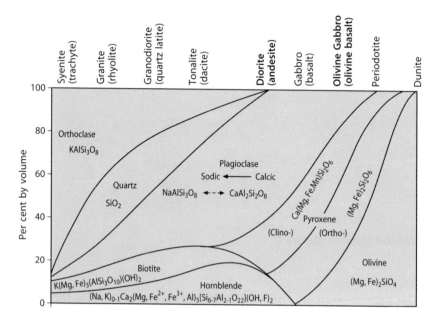

Fig. 2-1. General classification of igneous rocks showing principal minerals and their chemical compositions (modified from Mason and Moore, 1982)

Figure 2-1 identifies the principal coarse- and fine-textured igneous rocks and their mineral and chemical components.

During the crystallization of a magma, a process of chemical differentiation is active. As the temperature/pressure conditions in a magma change during cooling, minerals crystallize according to their stability fields. Different minerals will precipitate from a magma at limited ranges of temperature/pressure/chemical composition conditions. As minerals crystallize, magma mass decreases. The magma is depleted in the chemical elements that comprise the crystallizing minerals but the remaining magma is enriched in the elements that do not enter into those minerals. This is denominated differentiation. Although some potentially toxic metals are found in trace amounts in rock-forming minerals, they mainly concentrate in the ever decreasing volume/mass of residual magma. A few potentially toxic metals form their own mineral or may be an important component of a rock-forming mineral and are depleted in a crystallizing magma. These include Cr which crystallizes as the mineral chromite ($FeCr_2O_4$), and Ni which enters the rock-forming mineral forsterite ($Mg_2[Ni]SiO_4$) as a substitute for the element Mg, both during the early stages of differentiation.

As crystallization proceeds and differentiation intensifies, metal concentrations may increase to the degree that some can precipitate as their own mineral (e.g., U as uraninite, Be as beryl) or be hosted in late stage-forming accessory minerals. An example is the accessory mineral zircon which can accept elements such as U, Th and rare earth elements into its crystals. Some potentially toxic metals attain concentrations and reach conditions in a residual magma to crystallize as an accessory mineral. Accessory minerals comprise only a few percent of an igneous rock but may contain important metals. The Ti mineral rutile (TiO_2) is an example of this. Titanium can also bond with Fe and O to form the common accessory mineral ilmenite ($FeTiO_3$). Most metals tend to concentrate in hot residual (hydrothermal) fluids during the latter stages of magma differentiation. They may be injected or infiltrate into enclosing rock and precipitate as ore minerals as temperatures drop and chemical reactions take place between the hydrothermal fluid and the rock. These include Hg as cinnabar (HgS), As as arsenopyrite ($FeAsS$), Pb as galena (PbS), Zn as sphalerite (ZnS), Cu as chalcopyrite ($CuFeS_2$), Mo in molybdenite (MoS_2), Fe as pyrite (FeS_2) and U as uraninite (UO_2). Still others may be hosted in an ore mineral. Examples of this are Cd which substitutes in part for Zn in sphalerite ($Zn[Cd]S$, and As which can accompany Fe in the mineral pyrite ($Fe[As]S_2$). Ore most often occurs as assemblages of several minerals so that the smelting and processing of ore to recover one or more metals can result in the release of others into the environment.

Weathering and Soils

The primary igneous rocks and associated ore minerals undergo other
processes that result in the mobilization and redistribution of mineral
and chemical components and a reorganization and recrystallization
into new minerals. Weathering is the process that causes the disintegration and decomposition of preexisting rock as it interfaces with air,
water and organisms (e.g., bacteria) at or close to the earth's surface.
Some mineral components of rock are essentially insoluble and others
soluble. Soluble minerals decompose by interacting with carbonic acid
(water + CO_2 from the atmosphere and biological activity) and their
chemical elements go into solution. Relatively insoluble minerals such
as quartz (SiO_2) disintegrate from the rock mass as solid particles. The
total process is called weathering and the end result is soil formation.
Part of the soluble chemical elements enters into groundwater or runs
off into a stream and are moved out of the immediate environment.
Another part reacts to forms new minerals stable under ambient conditions. The combination of new minerals and solid particles from the
disintegration process forms soils.

As soils develop during weathering, the mass of matter differ-
entiates into distinct layers (horizons) that comprise a soil profile
(Figure 2-2). Each horizon contains different types or proportions of
earth materials. This is the result of processes active in the developing
profile. For example, an A horizon is composed of mineral and organic
phases underlying a mainly organic horizon of decomposing and
decomposed vegetation (humus). The A horizon is subject to leaching
as rainwater infiltrates downward. This is referred to as the zone of
eluviation as the seeping waters move clay-size particulates and mobilized chemical elements (e.g., Fe, Ca, Mg and potentially toxic metals)
into the underlying B horizon. This is the zone of illuviation or accumulation of eluviated matter. Clay minerals and Fe oxyhydroxides
dominate the B horizon and adsorb potentially toxic metals mobilized
into it, possibly building up contaminant concentrations. The B zone

Soil horizons

O - Composed mostly of organic matter including decomposing leaves, twigs, etc.

A - Composed of mineral and organic matter. Zone of leaching (eluviation by rain/groundwater or other fluids moving clay and dissolved elements (e.g., Fe, Ca and Mg) to the B horizon. Lower part of the A horizon is the E horizon.

B - Composed of earth materials enriched in clay, Fe oxyhdroxides, $CaCO_3$, and other constituents leached from the overlying horizons. This is the horizon of accumulation (illuviation).

C - Composed of partially weathered (disintegrated-decomposed) parent rock.

Fig. 2-2. Soil horizons that develop in a temperate humid climate (modified from Drever, 1997)

is the main source for plant nutrients tapped by root systems. Any potentially toxic metals in the horizon may be translocated into the vegetation together with nutrients or as essential micronutrients. The C horizon underlies the B zone and is composed of partially altered parent rock from which the soil is continually forming. This is underlain by unaltered parent rock.

Soils may take 100s to 1000s of years to form depending mainly on temperature and precipitation and their influence on vegetation and biological activity but also on the parent rock, topography, and time. These factors determine the type of soil that develops and define ecological zones. Figure 2-3 illustrates the relationship between climate (temperature, precipitation, humidity), soil formation, and vegetation adapted to such zones. Figure 2-4 underlines the parallel relationship that exists between latitudinal and altitudinal distribution of vegetation and ecosystems which will also influence the type of soil that will form. The natural chemical composition of soils mirrors, to a good degree, the chemistry of the rock from which they originated. Some vegetation will incorporate chemical elements into their tissue in

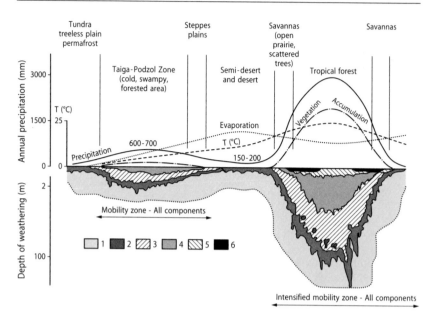

Fig. 2-3. The relationship between soil formation, climate (temperature and precipitation), latitude, and depth of weathering (from Strakhov, 1960). Symbols sequence reflects greater intensity of weathering: 1 = unweathered crust; 2 = grus zone, little chemical change; 3 = hydromica-montmorillonite-beidellite zone; 4 = kaolinite zone; 5 = Fe oxyhydroxide, Al_2O_3 zone; 6 = Fe_2O_3 - Al_2O_3 laterite/bauxite zone

proportion to their concentrations in soils and hence reflect soil (and rock) chemistry. This is the basis for biogeochemical exploration for mineral deposits and establishes natural (baseline) concentrations for species at sampling locations. Because rocks within a given class or among different classes have varying chemical compositions (Table 2-1), soils can have markedly varying chemical compositions. For example, basalt is enriched in several potentially toxic metals (e.g., Cr, Co and Ni) with respect to granite. Thus, a soil formed from basalts can be expected to contain higher concentrations of Cr, Ni and Co than a soil formed from granite. Similarly, ocean clay sediment is enriched in As, Co, Cu, Mo, Ni, Pb and Zn with respect to the igneous

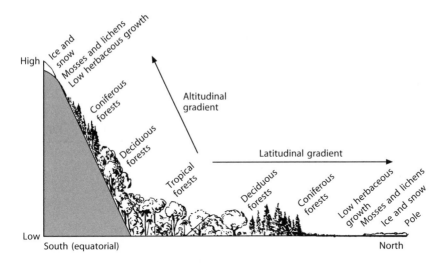

Fig. 2-4. The similarity between latitudinal and elevation distributions of vegetation and hence ecosystems (after Simpson and Beck, 1965). The soil that forms will be influenced by this geographic-altitudinal factor

rocks represented in Table 2-1. Soils formed on marine shale, the rock equivalent of ocean clay can be expected to be enriched in an assemblage of these metals. Thus baseline or natural background concentrations of metals in soils can be notably different from one area to another. Because of this, the concept of using global, national or regional element concentration averages as background values and hence remediation targets of contaminated soils is not valid. Extreme weathering conditions can remove nutrients from a soil so that it will not sustain vegetation growth and is no longer a viable soil. In such a case, if most chemical elements are lost during decomposition, aluminum oxide ($Al_2O_3 \cdot 3H_2O$, bauxite) and iron hydroxide (FeOOH, goethite) may remain as the residual product called laterite. Laterite that originates from olivine basalts or gabbros are often mined for their Ni and Co contents. If the iron hydroxide is removed, bauxite, the principal ore of Al remains.

Table 2-1. The average natural contents in selected rocks of metals/metalloids that are potentially toxic to humans and other life forms. Compilation from several sources. Values in ppm unless otherwise noted

Metal	Earth Materials						
	Granite	Basalt	Shale	Ocean Clay	Lime-stone	Deep-Sea Carbonate	Streams (ppb)
Al%	7.2	8.2	8.0	8.4	0.42	2.0	50
As	2	2.2	13	13	1	1	2
Be	3	0.7	3	2.6	0.X	0.X	0.01
Cd	0.13	0.21	0.3	0.03	0.3	0.0X	0.01
Co	4	47	20	74	0.1	7	0.1
Cr	10	185	100	90	11	11	1
Cu	20	94	50	250	4	30	7
Fe%	1.42	8.6	5.1	6.50	0.38	0.9	40
Hg ppb	0.03	0.09	0.4	0.03	0.04	0.0X	0.07
Mn%	0.045	0.18	0.09	0.07	0.11	0.1	7
Mo	1	1.5	2.627	0.4	3		0.6
Ni	10	145	60	230	20	30	0.3
Pb	17	7	20	30	9	9	1
Sb	0.22	0.6	1.5	1	0.2	0.15	0.07
Sc	7	27	16	19	1	2	0.004
Se	0.05	0.05	0.6	0.17	0.08	0.17	0.06
Sn	3	1.5	6	1.5	0.X	0.X	0.04
Ti%	0.12	1.14	0.60	0.46	0.04	0.08	3
Tl	2.3	0.21	1.4	0.8	0.0X	0.16	–
V	50	225	140	120	20	20	0.9
Zn	50	118	85	200	20	35	20

Assumed Shale Equivalent (volatile-free, carbonate-free basis): McLennan and Murray (1999). Average Pelagic (Ocean) Clay: McLennan and Murray (1999). Stream Water: Taylor and McLennan (1985). Basalt: Average Turekian and Wedepohl (1961); Vinogradov (1962). Granite: Low Ca, Turekian and Wedepohl (1961). Soil and Natural Vegetation: Connor and Shacklette (1975); Shacklette and Boerngen (1984); Brownlow (1996).

Table 2-1 (continued)

	Cultivated Soil	Uncultivated Soil	Vegetation Ash	
			Natural	On Mineralized Terrain
Al%		1.1–6.5	0.1–3.9	
As	5.5–12	6.7–13		
Be	1–1.2	0.76–1.3		
Cd		0.1–0.7	0.95–20	
Co	1.3–10	1–14	0.65–400	>50
Cr	15–70	11–78	2.2–22	
Cu	9.9–39	8.7–33	50–270	50–60 to 100–200 Sometimes >1000
Fe%	1.4–2.8	0.47–4.3	0.08–0.93	
Hg ppb	30–69	45–160		
Mn%	0.099–0.74	0.006–0.11	0.05–1.4	
Mo		0.2–5	0.76–7.6	
Ni	1.8–18	4.4–23	0.81–130	>100 Sometimes 2000
Pb	2.6–27	2.6–25	24–480	
Sb		2.0		
Sc	2.8–9	2.1–13		
Se	0.28–0.74	0.27–0.73	0.01–0.42	
Sn		3–10		
Ti%	0.17–0.40	0.17–0.66	0.07–0.12	
V	20–93	15–110	2.6–23	
Zn	37–68	25–67	170–1800	300 Grasses 500–1000 Wood

Sedimentary Rocks

Disintegrated particles and soil are sediments. Erosion by running water, wind, and ice can pick up sediment and transport it to terrestrial and oceanic basins where it deposits. As new sediment accumulates, fluids or moisture is squeezed out of the older underlying sediments by compaction. Cements can precipitate and bind particles together over time to form a sedimentary rock. Shells from dead organisms that lived in the basins accumulate there as well. The shells are comprised of the minerals calcite and/or aragonite (both comprised of $CaCO_3$ but with different crystal structure) precipitated by the life forms which extracted the decomposition weathering products Ca^{2+} and the CO_3^{2-}, from basin waters. In unique environments where evaporation exceeds inflow of water with soluble products and supersaturation occurs (e.g., the Red Sea), chemical sediments comprised of the minerals halite (NaCl, salt), gypsum ($CaSO_4 \cdot 2H_2O$, plaster of paris), calcium carbonate ($CaCO_3$), phosphorite ($Ca_3(PO_4)_2$), borate salts (BO_3^-) and goethite (FeOOH) can precipitate directly in ocean/sea or land-locked basins. Sedimentary rocks are classified according to the dominant size of the bound particles and their chemical composition. Table 2-2 gives a classification of sedimentary rocks and illustrates their relation to sediments. Sedimentary rocks differ from other rocks because they have the properties of porosity (the ability to hold fluid) and permeability (the ability to transmit fluid). They are the aquifers for water that sustains living populations and about two-thirds of the global agricultural productivity. They contain oil and natural gas and originate coal-bearing strata thus satisfying about 80 % of global energy needs. Because of their porosity and permeability, where sedimentary rocks are invaded by ore-metal bearing hydrothermal fluids, important ore deposits of many potentially toxic metals (e.g., Pb, Zn and U) can form.

Table 2-2. General classification of sediments and sedimentary rocks

Sediment	Sedimentary Rock	Size and/or Composition
Gravel	Conglomerate	> 2 mm rounded rock and/or mineral detritus; in rock this is cemented by silica (SiO_2), calcite ($CaCO_3$), or iron oxides
Sand	Sandstone	1/16-2 mm particles, mainly quartz, cemented as above
Silt	Siltstone	1/256-1/16 mm particles as given above with cementation as above or by clay-size matrix
Clay	Shale	< 1/256 mm particles as weathering decomposition products of feldspars and other minerals; lithification by compaction and cementation
Accumulated Shells Lime Mud	Limestone	Rock of $CaCO_3$ from shell remains or from chemical precipitation by evaporation
	Dolostone	Limestone or lime mud altered by interaction of Mg-rich waters to the mineral dolomite $CaMg(CO_3)_2$
	Gyprock	Precipitate of $CaSO_4 \cdot 2H_2O$ (gypsum) from evaporation
	Rock Salt	Precipitate of NaCl (halite) from evaporation
	Coal	Accumulation of vegetation to form peat, lignite and bituminous coal under increasing compaction, and anthracite with increasing heat and pressure

Metamorphic Rocks

Metamorphic rocks comprise the third major rock class. These derive from any pre-existing rock (igneous, sedimentary or metamorphic) when it is brought to temperature and pressure conditions that are greater than that under which the pre-existing rock lithified. The chemistry of metamorphic rocks may reflect the chemistry of the rock from which it developed. Under high temperature/pressure conditions chemically active fluids may be generated. There is a metamorphosis when rock-forming minerals outside their stability fields reorganize and recrystallize to other rock-forming minerals that are stable under the new temperature/pressure and chemical conditions. Because high temperatures are reached and fluid phases are present, juxtaposition to enclosing rock at ambient temperatures results in physical-chemical reactions where stable minerals develop in a contact metamorphic zone. Important metal-bearing ore deposits originate during contact metamorphism. Metamorphic rocks are classified into different types based on macro- and micro-foliation and layering (results from directed pressure during metamorphism), texture (crystal size) and mineral composition.

The Rock Cycle

The relationship between magma, igneous, sedimentary and metamorphic rocks, and soils is given by the rock cycle (Figure 2-5). Two points previously cited are reemphasized in Figure 2-5. First, soils can develop from the weathering of any class of rock. Second, any rock may undergo metamorphism when it reaches equilibrium at temperature and/or pressure conditions greater than the ambient conditions under which they formed and where chemically active fluids affect the equilibrium between mineral/rock phases.

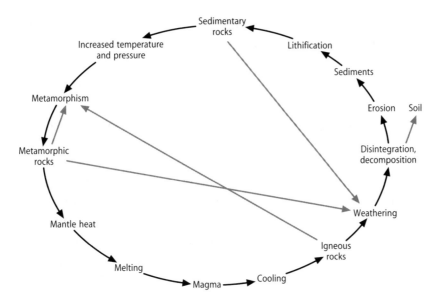

Fig. 2-5. The rock cycle, products and important conditions

Water

The natural chemistry of water on land is influenced mainly by the chemical composition of the rocks (soils) the water flows on or flows through, whether stream and river water, springs and aquifer water, or ponds and lakes. The waters take on the chemistry of rock-forming minerals and any metal sulfide minerals in the rocks, especially pyrite, FeS_2. This will be influenced in turn by the physical-chemical-biological characteristics of the waters such as temperature, pH, redox potential, adsorption or desorption from inorganic and organic suspended or bottom sediment, cation exchange, dilution-evaporation, and life forms present. Thus, water that flows over or through limestone, a rock made of $CaCO_3$, will develop a pH of about 8.1. However, waters flowing over granite, [mainly quartz (SiO_2) and feldspars ($KAlSi_3O_8$, $NaAlSi_3O_8 - CaAl_2Si_2O_8$)] or through fractures in it will have a pH of about 6. If either of these rocks contains pyrite, omnipresent in sulfide ore deposits, oxidation of the mineral will result in the generation of acid waters that can affect the pH of the greater water mass. Where limestone is the host rock, mixing with acid water from pyrite oxidation can cause the pH to drop from 8.1 to 7.0. This was found to be the case in streams associated with Mississippi-Valley type stratabound sulfide ores in Virginia (Siegel, 1990), but with no apparent impact on the fluvial environment. Where granite is involved, added acidity from pyrite oxidation may cause the pH to drop to 5 or lower and cause a serious acid rock drainage problem for life in an ecosystem. In the previous chapter, Figure 1-1 related, in a dramatic example, the response of several aquatic life forms to pH changes (from acid rain, from mineral-water reactions) in the ecosystem. As the pH drops, organisms in an ecosystem either move out or do not survive.

Atmosphere

The atmosphere carries potentially toxic metals as gases, aerosols and particulates that precipitate and become part of the earth's surface/near-surface ecosystems. These metals can originate naturally from volcanic emissions, geysers, thermal springs, forest fires, and volatilization of metals such as Hg and Se from ore deposits and soils. They are also introduced into the atmosphere by high winds that raise metal-bearing fine-size particles into the atmosphere, and as micrometeorites < 1 mm in diameter that enter the atmosphere and ultimately reach the earth at a rate of at least 6000 tons annually (Ponchet, 2000). This is far less than the 30000 or more tons annually estimated by Mason and Moore (1982). The addition of metals to the earth's surface from atmospheric sources and their uneven geographic distribution varies with meteorological factors (e.g., velocity and directional properties of wind driven air masses). The distribution also varies with a metal's residence time in the atmosphere. For example, Hg may be carried in arctic air masses from 3 months to 3 years (AMAP, 1997). One result of this is that baseline or natural concentrations of metals from these sources may vary significantly from one study area to another.

From Human Activities: Inadvertently and Knowingly

Potentially toxic metals can be released into a living environment inadvertently by human activities that affect physical and chemical conditions in a natural setting. More commonly, they have been introduced knowingly into ecosystems from industrial and agricultural operations in effluent streams, as chimney emissions and in run-off without attention to treatment or capture previous to release. Table 2-3 gives examples of the multiple uses of potentially toxic metals that can ultimately result in their being released or mobilized into an environment.

In recent years and as a result of legislation in many countries requiring capture and treatment of pollutants, loading of ecosystems with heavy metals has lessened considerably. However, this is not global. A factor which influences a nation's dedication to reestablishing and maintaining healthy and sustainable ecosystems is the enforcement of existing laws. In the case of atmospheric emissions, there are laws permitting the sale of unused allowable pollutant release by one company to another. This does not alleviate atmospheric pollution. Laws in some countries allow for a "polluter pays" principle but this is not environmentally sound for two reasons. First, paying to pollute and passing the cost on to consumers only exacerbates environmental problems and results in economic (higher costs) problems. Second, the "paying polluters" can (greatly) understate the amounts of pollutant release to an environment sometimes with the knowledge of

Table 2-3. Many and varied uses of potentially toxic metals through which they can be introduced into a living environment. Compilation from several sources

As: Additive to animal feed, wood preservative (copper chrome arsenate), special glasses, ceramics, pesticides, insecticides, herbicides, fungicides, rodenticides, algicides, sheep dip, electronic components (e.g., gallium arsenate semiconductor, integrated circuit, diodes, infra-red detector and laser technology), non-ferrous smelters, metallurgy, coal-fired and geothermal electrical generating facilities, textile and tanning, pigments and anti-fouling paints, light filter, fireworks, veterinary medicine

Be: Alloy (with Cu), electrical insulators in power transistors, moderator or neutron deflector in nuclear reactors

Cd: Ni/Cd batteries, pigments, anti-corrosive coatings of metals, plastic stabilizers, alloys, coal combustion, neutron absorber in nuclear reactors

Co: Metallurgy (in superalloys), ceramics, glasses, paints

Cr: Manufacturing of ferro-alloys (special steels), plating operations, pigments, textiles and leather tanning, passivate the corrosion of cooling circuits, wood treatment, audio, video and data storage

Cu: Good conductor of heat and electricity, water pipes, roofing, kitchenware, chemicals and pharmaceutical equipment, pigment, alloys

Fe: Cast iron, wrought iron, steel, alloys, construction, transportation, machine-manufacturing

Hg: Extracting of metals by amalgamation, mobile cathode in the chloralkali cell for the production of NaOH and Cl_2 from brine, electrical and measuring apparatus, fungicides, catalysts, pharmaceuticals, dental fillings, scientific instruments, rectifiers, oscillators, electrodes, mercury vapour lamps, X-Ray tubes, solders

Mn: Production of ferromanganese steels, electrolytic manganese dioxide for use in batteries, alloys, catalysts, fungicide, antiknock agent, pigments, dryers, wood preservatives, coating welding rods

Mo: Alloying element in steels, cast irons and non-ferrous metals, in chemicals such as catalysts and dyes, lubricants, corrosion inhibitors, flame retardants, smoke repressants, electroplating

Table 2-3 (continued)

Ni: As an alloy in the steel industry, electroplating, Ni/Cd batteries, arc-welding rods, pigments for paints and ceramics, surgical and dental prostheses, molds for ceramic and glass containers, computer components, catalysts

Pb: Antiknock agents, tetramethyllead, lead-acid batteries, pigments, glassware and ceramics, plastics, in alloys, sheets, cable sheathings, solder, ordinance, pipes or tubing

Sb: Type-metal alloy (with lead to prevent corrosion), in electrical applications, Britannia metal, pewter, Queen's metal, Sterline, in primers and tracer cells in munitions manufacture, semiconductor, flameproof pigments and glass, medicines for parasitic diseases, as a nauseant, as an expectorant, combustion of fossil fuels

Se: In the glass industry, semiconductors, thermoelements, photoelectric and photo cells and xerographic materials, inorganic pigments, rubber production, stainless steel, lubricants, dandruff treatment.

Sn: Tin-plated steel, brasses, bronzes, pewter, dental amalgam, stabilizers, catalysts, pesticides

Ti: For white pigments (TiO_2), as a UV-filtering agent (suncream), as a nucleation agent for glass ceramics, as Ti alloy in aeronautics

Tl: Used for alloys (with Pb, Ag or Au) with special properties, in the electronics industry, for infrared optical systems, as a catalyst, deep temperature thermometers, low melting glasses, semiconductors, supraconductors

V: Steel production, in alloys, catalyst

Zn: Zinc alloys (bronze, brass), anti-corrosion coating, batteries, cans, PVC stabilizers, precipitating Au from cyanide solution, in medicines and chemicals, rubber industry, paints, soldering and welding fluxes

enforcement officials who accept monetary or material bribes. Strict monitoring systems must be in place and their results interpreted within existing laws in order to preserve environmental sustainability.

Irrigation

In some cases, normal human activities trigger unwitting releases of pollutant metals into an environment. This may be termed "the law of unintended consequences". An excellent example of this is related to the use of aquifer water for large scale agricultural development in West Bengal, India and East Bangladesh. Before large agricultural development began, a few hundred village wells were drawn on steadily with a relatively good balance between discharge and recharge thus maintaining the level of the water table and serving the needs of the large regional populations. During development more than 20,000 wells were sunk and heavily drawn on during the growing season. This lowered the water table seasonally and aerated strata comprising the aquifer which were not exposed to oxidation before the water demand increased. The aerated strata contain the mineral pyrite (FeS_2) which has As substituting for part of the Fe. Oxidation and decomposition of the pyrite released As into the aquifer where it formed the very toxic arsenite complex. About 200,000 people were diagnosed with arsenic poisoning from drinking and cooking with the water, and eating foodstuffs grown with the water. Many of them were seriously and irreversibly ill from various maladies including skin cancer. Another 10-20 million people are considered to be at high risk unless an alternate water supply is developed (Bagla and Kaiser, 1996; Dipankar et al., 1996; Nickson et al., 1998; Nickson et al., 2000; Islam et al., 2000).

Irrigation can result in waters containing one or more metals that have leached out of soils. If these waters discharge into surface or aquifer systems, the metal(s) may, over time, load the waters to pollutant levels. The water quality downflow may be unsuitable for human needs (e.g., for drinking water, for agriculture). This is the case with Colorado River waters that are used for irrigation in Arizona and California. The waters moving out of the irrigation system have a high Se content. This was not foreseen and made the water unuseable when it returned to the river. The problem had international implications since the end users were Mexican farmers. The United States resolved the problem by installing a treatment plant at Yuma, Arizona, where waters were brought to a useable quality before being returned to the river that flowed on to Mexico. Irrigation with waste waters can cause similar problems of a build up of mobile potentially toxic metals in the soils and in surface runoff. Although salinization of soils is not a heavy metals problem, it results from irrigation where there is poor drainage. The buildup of salts at the roots will diminish the productivity of soils with time unless flush/drainage systems are installed.

Biocide Runoff or Infiltration into Soils and Aquifers

Fungicides, rodenticides and herbicides are used to protect crops but at the same time enter the soil. Soil amendments are used to increase productivity by supplying additional nutrients (fertilizers) to crops or by changing soil conditions (e.g., pH) to make nutrients more bioavailable to them. Besides organic chemical contaminants, the biocides may contain potentially toxic metals such as As and Hg. These metals can further intrude an ecosystem from soil by water infiltration into unconfined aquifers or through runoff to fluvial systems. Soil amendments derived from sewage sludge, manure and dredged sediment from polluted harbors or rivers may contain a broad spectrum of potentially toxic metals (e.g., As, Cd, Hg, Se, Mo and Co). These can also intrude an ecosystem if they are mobilized out of a soil into vegetation, the atmosphere, or aquifer and/or surface waters.

Table 2-4. Mobilization of potentially toxic metals during dredging. Concentrations in ppm except Hg in ppb (after Darby et al., 1986)

PTM	Channel sediment pore water (a)	River water conc. (b)	a/b	Effluent at man-created marsh		
				Expected	Measured	% Change
Zn	0.12	0.052	2	0.065	5.30	+ 8069
Cu	0.012	0.004	3	0.0055	0.051	+ 827
Pb	0.077	0.002	38	0.016	0.142	+ 788
Cd	0.009	0.001	9	0.0025	0.019	+ 660
Ni	0.054	0.001	54	0.011	0.035	+ 218
Hg	3.2	0.26	12	0.82	2.0	+ 144
Mn	6.94	0.03	230	1.34	1.19	− 11
Fe	57.3	0.26	220	11.12	6.01	− 46

Sewage sludge, dredged sediment from polluted rivers, and manure used as soil amendments can also increase concentrations of potentially toxic metals in a soil ecosystem. Before recovery for treatment and use on fields, the metals from these sources are generally in the reduced state and most are initially relatively immobile. However, once the metals-bearing solids are removed from a reducing environment and deposited in an oxidizing one, many potentially toxic metals are readily mobilized and become bioavailable to crops. Table 2-4 illustrates the changes in mobilization of several potentially toxic metals that occur as a result of dredging sediments from a reduced environment and depositing them in an oxidizing environment. As expected the elements Fe and Mn precipitate as oxyhydroxides in an oxiding environment and are immobilized. However, the other elements analysed show great to significant enhancements in their mobility in the oxidizing environment from more than 8000 % for Zn to 144 % for Hg. As is the case with metals added to soils from biocides, metals from soil amendments can infiltrate aquifers or run off into surface waters.

Wastes from the cultivation of food animals and poultry can introduce potentially toxic metals (and pathogens) into soil systems. Seepage through these wastes into soils and runoff into drainage basins can load an environment with their contents of mobilized metals. In the past Hg entered an environment as coatings to preserve seeds.

Anthropogenic – Human Activities Caused and Acknowledged

Mining

Many of the metals in ore deposits have low concentrations so that their extraction produces a great amount of waste rock. For example, if Cu is being mined and the tenor of ore is 1 % or less as is the case for most porphyry deposits, about 20 pounds of Cu could in theory be won from a ton of mined rock. This leaves more than 1980 pounds of waste rock to be disposed of. Globally, mining produces thousands of millions of tons annually of waste rock. No matter how selective the picking of the ore-bearing rock is, some ore remains in the waste rock together with accompanying minerals such as the iron sulfide (FeS_2), pyrite which is not recovered for mineral processing. The waste is commonly disposed of on the earth's surface in piles of rock called tailings or spoil piles. This is an oxidizing environment. The tailings are exposed to weathering in a two stage process which can cause environmental intrusions either singly or in union. One is the generation of acid mine drainage from the oxidation of pyrite in the waste rock. Acid mine drainage can be devastating to an ecosystem as evidenced by Figure 1-1. The second is the mobilization of potentially toxic metals in the surface environment stimulated by the oxidizing weathering environment under acidic conditions from pyrite oxidation.

If potentially toxic metals are mobilized, most often in acid mine drainage, grave health problems can develop over time for a user population through respiration, drinking and cooking waters, and through soils and waters to foods. This has been especially true for elements such as As, Cd, Cu, Hg and Pb historically and during modern times. Several cases of heavy metals pollution from mining and their negative health impacts are cited in various chapters of this text.

Mineral processing effluents that flow into steams and rivers can carry toxic heavy metals. Perhaps the best example of this is the extraction of gold from placer deposits using a Hg amalgam. Mercury is released to the river waters in many placer mining areas and becomes part of the water-sediment load. Gold adhered to Hg is extracted by burning the Hg off thus creating an atmospheric pollutant. The pollutant will return to the earth's soil and water systems as atmospheric deposition and be a threat to ecosystem inhabitants as the pollutant passes up the food web and bioaccumulates in agricultural and animal/fish food products. This has taken place in recent years in placer gold mining in the Amazon where there has been no enforcement of laws controlling the disposal or release of mining wastes such as Hg into water courses (Cleary et al., 1994; Thornton, 1996; Lacerda et al., 1995). Heavy metal pollutants and associated acids or mineral processing liquors released as effluents will damage soils they flow over and seep into and water bodies into which they flow. The case of the overflow of a dam containing cyanide from the heap-leach process used to win fine Au from spoil piles at Baia Mare, Roumania was described earlier. In addition to the fish kill caused by the cyanide release, heavy metals in the leachate liquor further contaminated rivers that had previously been contaminated with heavy metals. Any thought of dredging the polluted sediments runs a real risk of mobilizing heavy metals from the sediment into the water system (Table 2-5) as well as into soils if dredged matter is used to amend soils.

Table 2-5. Emissions of air pollutants in northern former USSR in tons/yr (modified from Pacyna, 1995)

Region	As	Cd	Cr	Mn	Ni	Pb	Sb	Se	V	Zn
The Urals	551	145	1386	1158	1615	9530	94	180	3000	3920
The Norilsk Area	246	26	31	28	935	742	25	33	130	262
The Kola Peninsula	165	29	122	106	645	745	23	226	122	180
The Pechora Basin	12	4	81	74	73	198	12	22	66	56
% of Total from Cu-Ni Production	88%	53%	<1%	<1%	46%	20%	51%	33%	NG	17%
Fossil Fuel Combustion		6%	18%	16%	16%	28%	75%	16%	34%	87%	5%
Steel and Iron		5%	21%	79%	76%	9%	4%	27%	30%	8%	75%

Emissions

The emissions from smelting of ores has released millions of tons of toxins into the atmosphere annually as gases, aerosols and particulates. Potentially toxic metals are part of the toxin load. Pacyna (1995) calculated the tonnages of selected metals that have been emitted annually from anthropogenic activities on the Kola Peninsula, in the Urals, in Siberia and in the Pechora Basin (Table 2-5). Copper-Ni production, fossil fuel combustion and steel and iron manufacture contribute more than 90% of the atmospheric loading in the European Arctic for the elements tabulated, except for Ni with 83%. The values given from this relatively restricted region are enormous but represent only small percentages of global emissions from anthropogenic sources. For example, As emissions from the northern former Soviet Union contribute 4.5% to global anthropogenic input while Cd, Pb and Zn contribute 2.4%, 3% and 2.4%, respectively. The masses of anthropogenic emissions to the atmosphere for these four metals are an order of magnitude greater than those from natural sources. For

Hg, natural sources (volcanos, hot springs, thermal vents, ores) exceed anthropogenic emissions by about two to one.

The damage to an ecosystem and its inhabitants from dry and wet atmospheric deposition can be devastating from gases such as sulfur dioxide which reacts with water in the atmosphere to form sulfuric acid and from metals such as Pb, As and Hg. The impact can be through respiration of the pollutants and from atmospheric deposition of the metals which become part of soils and terrestrial, estuarine and ocean waters and sediments through which they may access the food web. Acid rain, dominated by sulfuric acid, abets the mobilization and bioavailability of many heavy metals. Once in the food web, the metals have the potential to bioaccumulate in foodstuff along the way. Consumers of metals-contaminated foods can, over time, develop health problems as they themselves bioaccumulate toxins which prevent the normal functioning of body organs and biochemical processes.

Heavy metal-bearing emissions originate from the fossil fuel-fired electrical generating facilities and other industrial processes and from the use of fossil fuels and wood for home heating and cooking. Without chemical scrubbers and electrical precipitators in chimneys, the metals enter the atmosphere according to their volatility. They are transported by wind currents until brought to the earth's surface via dry deposition and precipitation. Most of the potentially toxic metals will be found in high concentrations in earth materials deposited and maintained under reducing conditions and containing significant concentrations of sulfur and organic matter (e.g., coal, oil, black shales). These are the Goldschmidt denominated "sulfur-loving" or chalcophile elements As, Cd, Co, Cu, Fe, Mo, Ni, Pb, Sb, Sc, Se, V, Zn and others not treated in this text. Combustion of coal and petroleum releases emissions of volatile metals and metal-bearing particulates through chimneys.

Effluents

Industrial effluents discharge heavy metals into water courses, ponds and lakes, lagoons and other wetlands, estuaries and oceans. Some of the metals load is transported to a sedimentary environment in solution, part is carried as suspended sediments (inorganic and organic), and some is moved as bedload. Metals in the first two phases are bioavailable and readily mobilized into a food web before they are deposited in a sediment. Metals in a bedload or surface sediment may be scavenged by bottom feeders and also enter a food web. However, once deposited, heavy metals in a sediment may be immobilized by the physical-chemical conditions of the bottom sediment. For example, As adsorbed to Fe oxyhydroxides is immobile as long as a necessary level of oxidation persists. Other heavy metals (e.g., Hg, Cd, Cu, Pb, Zn) are immobile at redox levels specific to them under the pH conditions of a sedimentary environment, especially in the presence of the sulfide ion. The mobilization or immobilization potential is much the same from within a sedimentary sequence as well as in surface sediments.

Specific industries generate known potentially toxic metals assemblages in their effluent streams (Table 2-6). For many industries, regulations for pretreatment of effluents are being drawn up and evaluated previous to presenting them for legislation (Table 2-7). However, this is a slow process because of industries' resistance and the political lobbying that is done to modify scientifically sound and preferred methodologies which can ultimately reduce costs to industry. In cases when environmental regulations are followed and enforced as regards effluent treatment for removal of contaminants before release to the ecosystem, the possibilities of damaging environmental intrusion are greatly reduced, although not entirely eliminated. Treatments are available for removing metals from effluents to meet environmental agencies' outflow concentration limits. This notwithstanding, if the costs for treatment are greater than can be recovered by product sales, an

Table 2-6. Some uses of potentially toxic metals through which they may be introduced into the environment directly as a result of industrial processes (e. g., via effluent discharge or emissions) or indirectly through runoff or waste disposal (dumping, landfill, incineration)

	As	Be	Cd	Co	Cr	Cu	Fe	Hg	Mn	Mo	Ni	Pb	Sb	Se	Sn	Ti	Tl	V	Zn
Alloys	•	•	•	•	•	•			•	•	•	•	•	•	•	•			•
Batteries and Electro/Chemical Cells			•					•	•		•	•		•					•
Biocides (Agriculture, Anti-Fouling)	•					•		•							•				
Ceramics and Glass				•								•			•		•		
Chemicals, Pharmaceuticals, Dental	•			•	•			•	•			•	•		•	•		•	•
Coatings (anti-corrosives)			•		•	•			•						•				•
Electrical Equipment and Apparatus	•		•			•		•				•	•						•
Fertilizers	•		•		•	•		•	•	•	•	•		•					•
Fossil Fuel Combustion (Electricity)	•		•		•			•	•		•	•	•	•			•	•	•
Mining, Smelting, Metallurgy	•	•	•	•	•	•	•	•	•		•	•							
Nuclear Reactor (Moderator, Absorber)		•																	
Paints and Pigments	•		•	•	•							•				•			•
Petroleum Refining	•		•	•							•	•						•	
Pipes, Sheets, Machinery						•	•					•							•
Plastics			•									•							
Pulp and Paper					•			•											
Rubber											•			•					•
Semi-conductors, Super-conductors	•					•								•					
Tanning and Textiles	•				•	•													
Wood Preservative Treatment	•				•	•						•	•		•	•	•		•

Table 2-7. Industries for which individual regulations for the predischarge treatment of effluents are being formulated by the USEPA (modified from Vesilind et al., 1994)

Metals Related Mining Foundaries Iron and steel workings Machinery Electroplating Copper working Aluminum working Coated coil manufacture Enameled products	*Energy Related* Thermal electrical generation (by coal, oil, natural gas, domestic wastes) Petroleum refining Batteries *Clothing* Textiles Leather tanning
Wood, Paper, Plastics, Synthetics Related Pulp and paper Timber processing Plastics/synthetics manufacture Gum and wood	*Chemical related* Pharmaceuticals Soaps and detergents Laundries Paint and ink manufacture Printing
Agriculture Livestock wastes and farming chemicals	Adhesives Explosives manufacture Rubber manufacture
Others Ceramics and glass manufacture Paving and roofing	

industry may have to shut down. Alternatively, investment in modernization of industrial processes and recyling or capturing heavy metals for later sale, will minimize much of the effluent metal output, be cost-effective in an acceptable time frame, and thus increase economic productivity.

Solid Waste Disposal – Industrial and Domestic

Solid wastes are generated worldwide in masses of thousands of millions of tons annually mainly from mining and agriculture but with the most toxic components input from industrial operations (Table 2-6) and electricity generation. The solid wastes such as manure from animal husbandry, coal ash from electricity generating or smelting facilities, and tailings from mining and metal extraction processes, contain high concentrations of potentially toxic metals (Table 2-6). Thus, they and metal-bearing wastes from domestic sources (e.g., batteries, tires, appliances, junked automobiles) that do not enter the recycling stream have to be disposed of, sometimes after treatment, in a secure repository so as to protect vulnerable environments. What should be done under the best conditions and what is actually done in many countries is quite different and slow to change. This is the result of the sheer magnitudes of the masses involved and the economics of finding, constructing and maintaining secure disposal sites.

For example, solid animal wastes can be used directly in soil conditioning. If so they pose a bacterial danger and can develop high concentrations of heavy metals which effectively concentrate in soils as wastes dry out and decompose. There can be a subsequent translocation and possible bioaccumulation in a field crop or forage. When animal wastes are gathered, dried out and used for fuel indoors, often with poor ventilation, the home atmosphere may be contaminated with contained volatile metals such as As, Hg, Se, Cd and Pb. The manure ash will have a greatly increased concentration of heavy metals and has to be properly disposed of. In some agricultural regions of India, fresh manure is collected, put in a specially valved 50 gallon drum and allowed to ferment and produce methane gas. The gas is drawn off and used as a supplemental fuel for cooking. The spent manure still has its heavy metal load and is used as a farm-land conditioner. Loading of soils excessively in this way poses a real threat to consumers if the food web is compromised by bioaccumulation of potentially toxic metals.

The fly ash from fossil fuel combustion presents major disposal problems because of the masses generated and the high contents of several potentially toxic metals in the fly ash. These metals were deposited and preserved with accumulating coal-forming matter in a reducing environment. If left exposed to weathering, the ash would yield heavy metals to leachate mobilized possibly in acid solution which could seep into an unconfined aquifers or move as runoff into surface drainage. In the former case this would lead to a long-term health hazard for users of the groundwater. In the latter case, there would be an immediate impact on the receiving ecosystem from both the heavy metals and possible low pH (acid) drainage. A recycling of the metals-bearing ash for useable products would require that they be immobilized in the fabricated materials under ambient conditions.

Tons of rock have to be removed during mining for coal and ore minerals. The rock has to be sorted, picked and/or treated to concentrate the commodity. For example, coal is separated from the non-productive rock by breaking and washing. The waste rock such as black shale has a depositional environment association with coal. Black shales deposit in reducing environments similar to coal and contain several potentially toxic metals such as As, Cd, Cu, Hg, Mo, Ni, Pb, V, Zn. These adsorb to the clay minerals that make up the shales and can also precipitate as sulfide minerals together with the iron sulfide, pyrite. If heavy metals-bearing waste rock is disposed of in exposed tailings piles in temperate or tropical environments, the rock will decompose by oxidation and yield metals to surface drainage and aquifers probably mobilized in low pH (acidic) discharge. Similarly, metals found in ore deposits are not all extracted during the mining or later mineral processing. Those disposed of in tailings piles under conditions cited above will yield acid mine or acid rock drainage and this will mobilize heavy metals out of the rock and waste ore into surface and subsurface environments to the detriment of the ecosystems there. In such cases, treatment of the drainage before it enters into sensitive ecosystems is the ideal way to cope with the problem but a pre-treatment protocol has not commonly been followed in the past. In more countries, this is now part of mine development plans that are being enforced in mining operations (e.g., in Australia, Canada, Chile and the United States).

Nonetheless, the most toxic and concentrated metal-bearing solid wastes should be disposed of in a secure disposal site. This would be one that is not affected by physical natural hazards such as earthquakes, volcanic activity, mass movements and flooding, and one that is contained and isolated from surface and subsurface water courses. A site design should allow for leachate collection and transfer to a treatment plant for cleansing of potentially toxic metals. Any such disposal site must be continally monitored by air sampling devices and by wells sunk outside its periphery or any containment barrier that may have been implaced.

CHAPTER 3

Heavy Metals Mobility/Immobility in Environmental Media

Chemical elements are mobilized by physical, chemical and biological vectors. Elements move in solution as cations, anions and ionic complexes. They incorporate into solid inorganic phases (e.g., sediment, suspended sediment, particulates from natural or anthropogenic emissions) or are absorbed/adsorbed by them. The same is true for solid, perhaps vital, organic phases (e.g., soft and hard parts of organisms, particulate organic carbon). In these modes the chemical elements are transported to depositional environments on land or in water bodies by water, wind and glacial ice following surface drainage, aquifer flowpaths, and wind driven water and atmospheric currents. Mobilized heavy metals in an element assemblage can be carried to an environment in concentrations significantly higher than natural levels in speciated forms. If these are bioavailable, they will be toxic to life forms if bioaccumulated over a period of time. When this scenario is met, the metals pose a threat to basic links in an ecosystem foodweb as well as to environmental niches and the ecosystem they comprise.

Controlling Parameters

Factors such as pH, redox potential, and temperature by themselves, but mainly in combination and often abetted by bacterial processes, affect the solubility, mobilization and precipitation/deposition of potentially toxic metals. These and several other factors in complex reactions determine the chemical forms (metal species) that are introduced to an environment. They also influence changes of metals species that may take place once equilibrium is established during interaction with an environment. Other important parameters that affect heavy metals mobility in an environment include soil/sediment textural heterogeneity (e.g., grain size), soil/sediment matrix composition (e.g., mineralogy, organic matter content), fluid or particle interaction with interstitial or overlying waters, and organism activity. These factors can vary greatly from one environment to another and even among niches in an ecosystem. This greatly limits generalizations on rules from which we can predict impacts that interactions of combined factors have on changes in mobility/bioavailability of metals. Nonetheless, for some parameters or combinations of them, predictions can be made and used in evaluations of the vitality and diversity of an ecosystem and its relation to heavy metals input.

Mobilization by physical/directional transport of media bearing potentially toxic metals will be considered first.

Physical Dispersion

The physical transport of solid matter by moving water leads to a geochemical fractionation based on size and specific gravity as gradient lessens, hydraulic energy drops, and sedimentation takes place. The fractionation may be evident in the sand size fraction of water-borne sediment where heavy minerals drop out of the transport mode as a function of their specific gravity. Heavy metals in minerals such as Sn from cassiterite (SnO_2), Cr from chromite ($FeCr_2O_4$), and Ti from rutile (TiO_2) are resistant to decomposition during weathering or contact with an aqueous oxidizing environment. Geographic distributions for metals in heavy, resistant minerals plot as high concentration strands in near-shore marine sedimentary deposits or as decaying contents downflow in fluvial deposits.

The fine-size clay fraction heavy metals component may also show linear or directional linear distributions in bottom sediments that reflect marine current flowpaths. However, as suspended sediments fall through a marine water column towards depositional environments they may move through subsurface and bottom currents that imprint their own directional flow signatures on heavy metals distribution and obscure the influence of surface currents. There can be a contaminant release into bottom currents from containerized wastes, 5 % of which rupture on impact when dumped in the ocean (MEDEA, 1997). Contaminant wastes can also be released into bottom currents by corrosion of containers. The heavy metal distribution as the metals react and precipitate will correspond to bottom current flow direction. This was suggested as the process that caused high value As concentrations in fine-size sediments from the St. Anna Trough in the Kara Sea where lewisite, an As-bearing (35 %) blistering agent used in chemical weapons may have been dumped (Siegel et al., 2000). In high latitudes, sediments trapped in sea ice (polynya) or adhered to sea ice together with phytoplankton, epontic ice algae and macrophytes will be moved by surface currents and deposit when the ice melts. This ice-rafting process can result in pods of high contents of heavy metals in glacial-marine bottom sediments that may not be linearly connected so as to mark current flowpaths.

Sulfide ore minerals such as sphalerite (ZnS), chalcopyrite ($CuFeS_2$) and associated non-ore minerals such as pyrite (FeS_2) decompose readily and rapidly under aqueous oxidizing conditions. They release heavy metals to environments along with acid producing hydrogen ions that can originate acid mine drainage and acid rock drainage.

This explains why sulfide minerals are rarely found in stream sediments. Thus, chemical reactions add a facet to the physical dispersion factor. Depending on specific ambient conditions, the mobilized metals may form new minerals (e.g. smithsonite, $ZnCO_3$ in a limestone/dolomite sequence), close to where decomposition occurs. Heavy metals released during oxidation mainly sorb to fine-size solids with charged substrates such as amorphous and crystalline iron and manganese oxyhydroxides, clay minerals (e.g., smectite/montmorillonite) and particulate organic matter. Concentration and mobilization of potentially toxic metals is abetted as fine-size sediment contacts mobilized metals during physical transport. For example, As is adsorbed onto Fe oxides phases and Pb sorbs to Mn oxide phases, whereas other potentially toxic metals such as Cd, Co, Cu, Mo, Ni, V and Zn sorb to smectites. (Table 3-1). Particulate organic carbon phases in the fine-size sediment fraction can carry high concentrations of Hg, Cu, Ni, V and Zn. However, the possibility exists that soil adsorption sites can become saturated and cause a release of H^+ so that pH decreases and element mobility increases in an aqueous phase. In environmental programs that evaluate the mobility and bioavailability of potentially toxic metals from a soil/sediment, selective extraction techniques assume great importance. These assess how the metals are bound among solid phases (e.g., as soluble forms, in exchangeable positions, in carbonate

Table 3-1. Potentially toxic metals normally found with secondary minerals or amorphous forms in soils/sediments (modified from Sposito, 1983

Mineral	Sorbed or precipitated metals
Fe oxides	V, Mn, Ni, Cu, Zn, Mo, As
Mn oxides	Fe, Co, Ni, Zn, Pb
Clay minerals	
Smectites aka	
Montmorillonites	Ti, V, Cr, Mn, Fe, Co, Ni, Cu, Zn, Pb
Illites	V, Ni, Co, Cr, Zn, Cu, Pb
Vermiculites	Ti, Mn, Fe

minerals, as easily reducible or moderately reducible phases, as sulfides, and in residual/resistant forms). This is discussed in Chapter 6. The strength of bonding is directly related to bioavailability.

Organisms can cause a physical redistribution of potentially toxic metals in lake and marine sediment sequences. To obtain needed nutrients, burrowing organisms disperse and redistribute metal-bearing sediment from the surface downward in a process called bioturbation. The depth of the redistribution and dispersion is 20 to 30 cm. For depositional environments where average rates of sedimentation are low, such as 0.1–0.6 mm annually, samples 20 to 30 cm deep should represent at the minimum sediment deposited at least 333 to 500 years ago. This is well before pre-global industrialization times so that potentially toxic metals in samples deeper than 30 cm should represent natural baseline concentrations. However, because large scale anthropogenic contamination of the environment began about 200 years ago, bioturbation likely affects sediment below that 200 year surface. Thus sediment deeper than 30 cm has to be used to determine natural baseline concentrations of target metals. In practice, bioturbation requires that samples depths of at least 40 cm be used in areas with sedimentation rates cited above to establish natural background levels of heavy metals. Where sedimentation rates are higher, baseline contents will be at greater depths.

A cultural or anthropogenic enrichment factor can be calculated by dividing the average or highest concentrations in the surface/near-surface bioturbated section by the average metal concentrations in the deeper, older sediment. However, the physical dispersion of heavy metals down into the sedimentary sequence dilutes what may have been higher surface concentrations. In such a case, a calculated cultural enrichment factor would be less than the true one.

Bioturbation that moves contaminated matter downward from a sediment/soil surface has two mobility/accessibility positive aspects. First, in aquatic sediments, it dilutes metal contaminants so that lesser concentrations are bioavailable to surface sediment bottom feeders. This makes less avaiable to a food web. Second, in irrigated fields, bioturbation can dilute surface/near-surface heavy metals concentrations and move metals down beyond the reach of food crop or animal forage root systems.

Mobilization in Aqueous Systems

In hydrogeologic sedimentary environments, some potentially toxic metals may be mobilized in solution as free cations (e.g., Cu^{2+}, Zn^{2+}, Co^{2+}, Hg^{2+}) or ionic complexes (e.g., $H_2AsO_3^-$, $H_2AsO_4^-$, $Cr_2O_7^{2-}$, $V_4O_9^{2-}$, $HgCl_4^{2-}$). As noted above, metals can be physically mobilized as species sorbed onto and transported by Fe and Mn oxyhydroxides, clay minerals and organic matter as fine-size (< 3.9 μm) particulates. Goldschmidt (1937) attributed the mobility responses among heavy metals to ionic charge similarities and ionic radii (Z/r) differences. According to Figure 3-1, Pb, Hg, Cd, Cu, Co, Fe and Zn can be found as soluble cations in sedimentary environments and the heavy metals Mo, As, Cr, Se and V can be present as anionic complexes in sedimentary environments. However, as will be demonstrated in the following text, this is a function of other influencing factors such as pH or Eh and sorption onto charged sites of fine-size solids. For example, the elements cited above and others such as Be, Sc, Ti and Sn are often present sorbed to unsatisfied charge sites on clay mineral hydrolysates. In solution and as sorbed species, cations, anions and/or anionic complexes of the heavy metals, are bioavailable and pose a risk to the health of an ecosystem.

An understanding of chemical mobilities and physical transport of heavy metals that results from active geologic processes can be used to indicate probable zones of provenance and from there to point sources within a provenance area. In a like manner, a knowledge of the mobility potential of toxic metals under existing environmental conditions and how this might change as natural or human-induced conditions change (with time) will assist remediation planning.

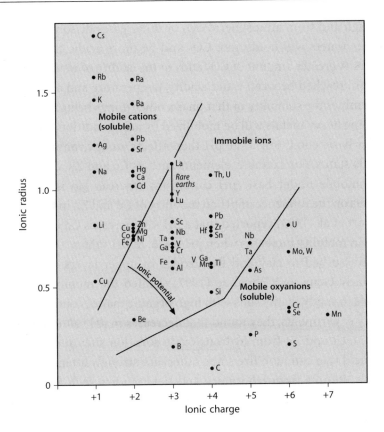

Fig. 3-1. The mobility of trace elements in relation to ionic charge and ionic radius (Z/r) and their reactions in sedimentary and hydrogeologic environments (after Siegel, 1992; based on the concept of Goldschmidt, 1937 and modified from Rose et al., 1979)

Hydrologic Mobility, Dispersion and Precipitation/Deposition

Temperature of water affects the solubility and hence the mobilization of chemical elements. At near neutral pH, solubility of chemical elements from their solid hosts generally increases with higher temperature. Natural waters develop pH values from the rocks they flow over or through and from atmospheric CO_2 or other gases dissolved in them. Colder waters will hold more CO_2 and be more acidic than warmer waters. A greater amount of CO_2 adds to the acidity of water. The equilibrium reached between water acidity, temperature and a solid phase determines the solubility of that phase, other things being equal.

Some heavy metals will be mobilized by acid conditions and others under basic conditions. The pH then affects an environment's heavy metals status. For example, elements such as Cu and Zn are essentially immobile under basic pH conditions whereas Mo is mobile. In studies on the influence of pH on mobility of Cd and Zn in loamy soils, Scokart et al. (1983) reported that a pH < 6 increases Cd mobility but that Zn mobility increases when pH is < 5. At pH values > 6 Cd and Zn remain sorbed to clay and free oxides in upper layers of the soils. Similarly Gong and Donahoe (1997) studied the mobility of heavy metals in sandy loamy soils with high organic matter content. In laboratory experiments, they found that decreases in pH values mobilized Cd, Cu, Cr and Zn from hydrous oxide sorption sites and concluded that acid rain can mobilize some otherwise strongly attenuated heavy metals. Results from laboratory and field studies such as these will certainly affect planning on the use of soil amendments. Likewise the results will indicate which methods being considered to remediate soils can best immobilize necessary nutrients or mobilize metals that could harm or interfer with the vital functions of organisms.

Different fluvial sediments can respond to changing environmental conditions by an enhancement or attenuation of mobilities of contained heavy metals. For example, Co, Mn and Ni were readily mobilized from sulfide-bearing fine-grained sediment into drainage as a result of a pH drop with oxidation of metal sulfides (Astrom, 1998). Copper was mobilized to a lesser degree and Cr and V had limited mobility. The mobility of heavy metals from sulfide-poor sediments that contain Fe and Mn oxides was pH dependent with mobility that followed the sequence Zn > Cu > Cd > Pb. In these sediments, As was immobilized by sorption onto Fe oxides. The order of mobility release of the metals is important to environmental research and planning.

The order of environmental concern about health impact on eco-system inhabitants depends on Pb and Cd concentrations in media at the site being investigated. It would be Pb > Cd > Cu > Zn or Cd > Pb > Cu > Zn.

Similarly redox potential (Eh) in an ecosystem can affect mobility in environmental systems independently but is more commonly coupled with pH in mobility studies. In flux-corer experiments, Van Ryssen et al. (1998) found that under renewed water oxic conditions outflow concentrations of Cd, Zn, Pb and Cu from sediments are at least 3–6 times higher than when water was not renewed. Redox changes in marine sedimentary deposits likewise can cause mobiliza-tion of heavy metals. Reducing conditions in interstitial waters within a sediment can mobilize some potentially toxic metals from their sulfide phases as free ions or ionic complexes. Among these are As (as arsenite, As^{3+}), Mn as Mn^{2+} and Mo as Mo^{3+}. These will move up-gradient towards the surface until a level of oxidation is reached that permits immobilizing reactions to occur. The immobilization can be by precipitation of a solid phase such a manganese oxyhydroxide or by sorption onto existing solid phases such as arsenate (As^{5+}) species onto iron oxyhydroxides. Likewise, arsenite can be mobilized under slightly reducing and acidic conditions in fluvial systems but will transform to arsenate and sorb to Fe oxyhydroxides if discharged into marine waters (basic and oxidizing). Conversely, ions of metals such as Zn, Cu and Hg may be mobile in oxidizing marine waters but when they are moved to a sulfide-containing reducing environment will precipitate as sulfide minerals. Figure 3-2 illustrates the relation-ship between environmental changes such as inundation, dessication and acidification, Eh and pH, and the mobility response of potentially toxic metals. For example, As mobility increases as Eh becomes more reducing and pH becomes more acidic whereas Cd, Hg and Pb become more mobile as Eh becomes more oxidizing and pH more acidic. The reversal of these parameters will cause immobilization of the elements cited. Data such as these are useful in deciding among remediation plans proposed for sites with different environmental conditions.

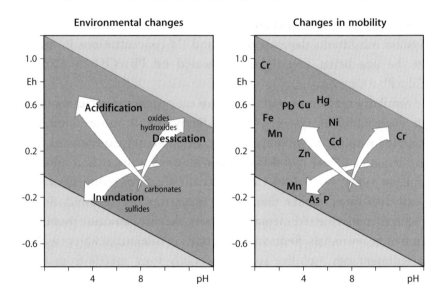

Fig. 3-2. pH-Eh (reduction-oxidation potential) diagrams illustrating the relationship between environmental changes and main mineral groups and the resulting trends of increasing mobilities of some potentially toxic metals (after Siegel, 1998; modified from Förstner, 1987 and Bourg, 1995)

Organisms in Mobility and Dispersion

Plants mobilize and translocate heavy metals from soils, waters and the atmosphere into edible and non-edible tissue. They provide a major pathway by which potentially toxic metals can enter a food chain in terrestial, lacustrine, estuarine and marine aquatic environments. The *in situ* terrestrial vegetation mobilizes and translocates nutrients and potentially toxic metals from soils into edible tissue. Vegetation may bioaccumulate the metals or exclude/limit their uptake. Feeders on this vegetation move whatever metals burden has bioaccumulated within a defined geographic range or may pass the metals up a food chain to organisms that can further biomagnify heavy metal concentrations. Toxicological effects to life along a food web are influenced by species and concentration of metals in an eco-system. Aquatic vegetation such as algae, phytoplankton and plankton can likewise bioaccumulate, exclude or limit uptake of potentially toxic metals from natural or contaminated waters. Planktonic organisms are physically mobilized by water currents, or moved, attached to ice in polar regions. During transport, the consumption of aquatic vegetation by fish and other life forms at at higher trophic levels moves bioaccumulated metals up a food chain where the process may continue. As prey are consumed by predators, the ultimate consumer may ingest concentrations of metals over time that become harmful to health. In general, humans are the ultimate consumers but in specialized environments such as the Arctic polar regions, the polar bear is at the top of the Arctic food web. Polar bear on Svalbard have moderately high hepatic concentrations of the very toxic methyl Hg (CH_3Hg^+) which is thought to originate from their main food source, the ringed seal. The seal, in turn, bioaccumulate metals from consumption of fish/shellfish at lower trophic levels and these in turn ultimately ingested potentially toxic metals from feeding on arctic ocean algae and phytoplankton. Thus, CH_3Hg^+ or other toxic species of metals ingested with fish or other prey will bioaccumulate in a predator and can be harmful to its health condition over time.

Upon death of an organism, recycling and the processes of immobilization/mobilization continue. Decomposing land vegetation (oxidizing conditions) releases nutrients and its contaminant load (e. g., metals and sulfur) to the surrounding environment. It may subsequently be incorporated in soils and waters from which contaminants originated and reenter a recycling mobilization pathway. Heavy metals from vegetation can be immobilized when it deposits and accumulates under reducing conditions (swamps) and is preserved. Under these conditions and in the presence of sulfide, potentially toxic chalcophile elements (e. g., As, Cd, Hg and others) are deposited together with the accumulated vegetation. As temperature and pressure increase with continued deposition above the vegetation-rich sediment there is an evolution to coal during millions of years. With combustion, heavy metals loads immobilized in coal during 10s to 100s of millions of years are mobilized once again as volatile emissions and particulates released into the atmosphere. Atmospheric precipitation and dry deposition put metals back into an ecosystem where mobilization/immobilization reactions are reinitiated. Combustion of wood, a principal fuel for cooking and heating in many areas of Africa, Southeast Asia and Latin America, adds significantly to atmospheric emissions and subsequent deposition of potentially toxic metals.

In like manner, upon the death of organisms other than plants, soft parts may decompose in an oxidizing water column and release contained metals to the environment. Soft parts may also be preserved in a reducing sedimentary environment and incorporate their contained heavy metals in sediment deposited there. Heavy metals such as Cu, Ni and V are found in petroleum and surely originate from organism soft parts that evolved into kerogen and then to petroleum under the range of temperature and pressure that comprise the oil or natural gas-forming window. Combustion of petroleum extracted from rocks 10s to 100s of millions of years old releases any contained heavy metals into the atmosphere from which they mobilize into an ecosystem via atmospheric deposition and long-term cycling begins once more.

Atmospheric Mobility and Dispersion Before Deposition

Emissions from the combustion of petroleum and coal at electrical generating facilities and industrial complexes such as smelters, as well as the coal and oil burned for home heating and cooking releases large amounts of metals such as As and Hg into the atmosphere (Chapter 2). Atmospheric deposition of such potentially toxic metals takes place downwind on land and in the oceans. The deposited metals are then moved by erosion (of soils), running water, and ocean currents to depositional sites. Atmospheric transport and deposition can be the principal source of many potentially toxic metals in some of the earth's fragile environments such as found in Arctic regions (Siegel et al., 2001 a, 2001 b).

Geochemical Barriers to Mobility

An understanding of the complexity involved in processes that stimulate the mobilization of heavy metals is the basis for the design of remediation methodologies. Mobilization of metals is a function of many factors some of which which can be effective for some metals but result in immobilization for others in an assemblage. The response of metals to these factors must be kept in mind during planning for remediation. In any project, it is essential to do a pilot study to test the efficiency of a selected cleanup technology. What is clear, however, is that pH and redox potential in ecosystems are principal driving forces in determining heavy metal mobility/immobility.

The relationship between pH, Eh and mobility is well illustrated in Table 3-2 in which Per'elman (1986) categorizes geochemical barriers to mobility in supergene (e.g., soil) environments. For example, when oxidizing waters interact with reducing (gley) conditions, Cu and Mo will be immobilized at any pH between < 3 to > 8.5 but Cr, Se and V in the same interacting waters will become immobile at pH values from 6.5 to > 8.5. Reducing (gley) waters that contain H_2S or interact with H_2S-bearing fluids, present a formidable barrier to many heavy metals mobile in the reducing environment (e.g., Cd, Pb, Hg). In the presence of H_2S, the metals exceed their solubility products, and precipitate as sulfide phases. The process is most effective when the pH barrier of the waters is < 3 – 6.5. When reducing (gley) waters interact with oxidizing waters, Fe and Mn oxyhydroxides will precipitate over a wide pH range.

Table 3-2. Examples of geochemical barrier conditions under which potentially toxic metals are immobilized in the supergene (e.g., soil) environment (modified from Per'elman, 1986)

	Strongly acidic pH < 3	Weakly acidic pH 3–6.5	Neutral and weakly basic pH 6.5–8.5	Strongly basic pH > 8.5
Oxidizing Waters				
Geochemical Barrier				
Oxidizing Reducing (H₂S)	Fe Cu, Hg, Pb, Cd, Sn, As, Sb, Mo	Fe, Mn, Co Mn, Co, Ni, Cu, Zn, Pb, Cd, Hg, Sn, Cr, Mo	Mn Cr, Mo, Se, V Mo, V	Cu, Zn, Cr, As
Reducing (gley)	Cu, Mo	Cu, Mo	Cu, Cr, Mo, Se, V	Cu, Cr, Mo, Se, V, As
Alkaline	Mn, Fe, Co, Ni, Cu, Zn, Pb, Cd, Hg, Cr, As	Co, Ni, Cu, Zn, Pb, Cd, Hg		
Adsorption	Sc, V, As	Zn, Cd, Ni, Co, Pb, Cu, V, Mo, As	Zn, (V, Mo, As)	V, Mo, As
Reducing Gley Waters				
Oxidizing	Fe	Fe, Mn, Co	(Fe), Mn Co	(Mn)
Reducing (H₂S)	Pb, Cd, Sn	Fe, Co, Ni, Pb, Cu, Zn, Cd, Hg	Fe, Co, Ni, Cu, Zn, Cd, Hg, (Mo)	Cu, Zn, Cd, Hg, Mn, (Fe, Co, Ni)

H_2S

Table 3-2 (continued)

	Strongly acidic pH < 3	Weakly acidic pH 3–6.5	Neutral and weakly basic pH 6.5–8.5	Strongly basic pH > 8.5
Reducing Gley Waters				
Reducing (gley)	Cu, Mo	Cu, Mo	Mo	Mo
Alkaline	Mn, Fe, Co, Ni, Cu, Zn, Pb, Cd, Hg, Cr, As	Mn, Fe, Co, Ni, Cu, Zn, Pb, Cd, Hg, Cr, As	Zn, Cd, Mn, Co, Ni	
Adsorption	Sc, V, As	Zn, Cd, Ni, Co, Pb, Cu, Fe, Mn	Zn	
Reducing H$_2$S Waters				
Oxidizing	Se, (Fe)	Se	Se	Se
Reducing (H$_2$S)				
Reducing (gley)				
Alkaline	Mn, Fe, Co, Ni, Zn, Pb, Cd, Cr, As			
Adsorption	Sc, V, As			

CHAPTER 4

Pathways, Cycles: Bioaccumulation, Impact on Living Ecosystems

Potentially toxic metals follow natural environmental pathways and cycles through the many ecosystems that provide for the very essence of life: water, food and waste disposal. To some degree they follow the geochemical cycles for the nutrients that sustain life (Figure 4-1): (O_2) supports respiratory metabolism; CO_2 is the source of carbon for photosynthesis; N_2 is an essential element of proteins; S is essential for protein and vitamin synthesis; and P is incorporated into many organic molecules and essential for metabolic energy use. Terrestrial, fluvial/lacustrine, estuarine and oceanic life forms can suffer short- or long-term perturbation if these pathways and cycles are intruded by natural events or impacted by human activities.

Disruptions of pathways can lead to mobilization of suspect metals and their bioavailability, and hence access to a food web. They can also immobilize metals and secure the natural integrity of a web. Immobilization could have a negative impact if it causes a deficiency of essential heavy metals micronutrients for ecosystem inhabitants.

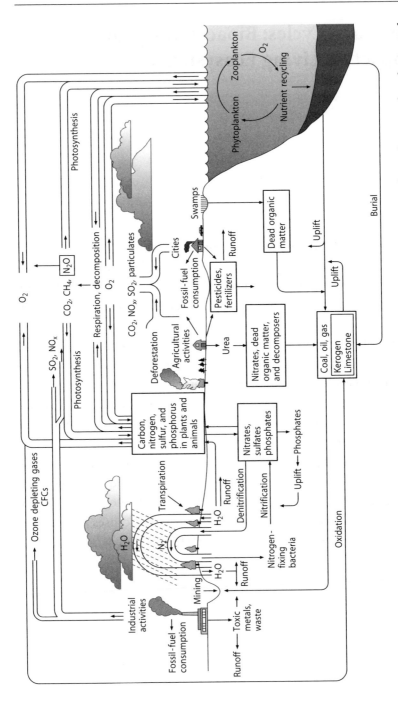

Fig. 4-1. A summary diagram of the biogeochemical cycles of the essential nutrients oxygen, carbon, nitrogen, phosphorus and sulfur (after Mackenzie, 1998)

PTMs Environmental Cycles and Pathways

The main pathways for access of potentially toxic metals to organism food webs are generalized in Figure 4-2. In natural systems, the metals originate from rocks, ore minerals (especially sulfide ores), and volcanoes and their associated fumaroles and thermal springs. Weathering releases metals during soil formation and they either become part of a soil or are transported to surface and/or aquifer waters. Depending on ambient physical, chemical and biological conditions, heavy metals in soils or waters may be bioavailable to a lesser or greater degree to a food web. The natural metals loading may be added to from anthropogenic sources. If the food web is accessed, various pathways can ultimately bring potentially toxic metals to human consumers and threaten their health. Similar pathway sequences can be generated for general or specific terrestrial, fluvial, estuarine and marine ecosystems or subsystems.

For example, in tropical rainforests, high rainfall and organic activity in warm/hot humid environments causes intense leaching and nutrient loss. If this continues without nutrient replacement, vegetation will not survive. However, in these ecosystems, the forest feeds on itself as decaying vegetation (e.g., fallen leaves, branches, trees and vines) replenishes nutrient supplies for living vegetation. During this cycle, there can be a buildup of heavy metals in the decaying organic matter (humus) and an accumulation in vegetation that is passed through to the rainforest food web. This can cause problems for organisms that bioaccumulate metals at higher trophic levels. On the positive side, living vegetation benefits an ecosystem by keeping soils stable against erosion. If the vegetation cycle is disrupted by a shortfall in nutrients and vegetation density diminishes, soils become susceptible to erosion and develop a greater deficit in available nutrients. This causes more vegetation to be lost and allows more erosion until a tropical rainforest ecosystem is decimated or lost.

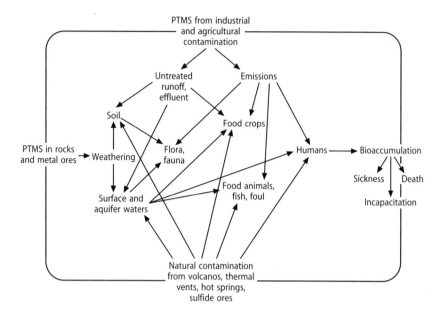

Fig. 4-2. Pathways of potentially toxic metals in the ecosystem that originate from natural and anthropogenic sources

Once in a food chain, heavy metals can bioaccumulate to dangerous levels in consumers. The risk to the health status of organisms becomes greater as metal concentrations increase along the food web. This can lead to decreases in biological productivity and hence sustainability of an ecosystem. Another consequence can be onslaught of disease(s) which may ultimately result in organism death. At the top of the food web, human (secondary) consumers can suffer the consequences of the ingestion of potentially toxic metals through water, food and atmospheric routes.

The atmosphere, hydrosphere (groundwater, surface waters), terrasphere (soils, sediments), and biosphere influences are inexorably and intimately interconnected in ecosystems. Linkages between them are complex but decipherable when making an assessment of an ecosystem health status.

Pathways

There are three principal pathways through which potentially toxic metals released from natural sources can access living things. The first is through the atmosphere directly, or indirectly through atmospheric deposition to soils and water. A second is through drinking water, water used for cooking, and irrigation waters for food crops. The third, fed from the atmosphere-water-soil complex, is through the food web.

Respiration

Atmospheric access is by respiration of natural and anthropogenic emissions of volatile and particulate heavy metals. For example, the natural release of Hg can be from volcanic activity (including fumaroles and hot springs), volatilization from rocks and soils, and evasion from water bodies. Atmospheric emissions from smelters and mineral processing, industrial manufacturing processes (e. g., chemicals, pulp and paper), electricity generating facilities, and heating, release scores of thousands of tons of potentially toxic metals (including Hg) to the atmosphere annually. For example, Pacyna and Keeler (1995) calculated that 1300 tons of Hg are emitted to the air annually mostly from Eurasia and North America. This is mainly from combustion of fossil fuels to produce electricity and heat with a major portion in a gaseous phase. Sixty to 80 tons of this is carried by air masses for deposition in the Arctic region and undoubtedly contributes to moderately high concentrations of hepatic Hg found in seabirds, seal, and polar bear from Svalbard and other Arctic locations. In Chapter 2 (Table 2-5) the release of thousands of tons of several metals (e. g., As, Ni, Cu, Cd, Zn) annually in smelter emissions from the Kola Peninsula, the Urals and Norilsk in Siberia was cited. This represents < 5 % of anthropogenic

Table 4-1. Global emissions (in tons) of some potentially toxic metals to the atmosphere in 1983. Modified from Pacyna (1995)

Global	As	Cd	Hg	Pb	Zn
Anthropogenic	20723	8322	3913	366347	145227
Natural	8598	1102	6614	20944	4409
% Anthropogenic contribution from northern former Soviet Union (Table 2-5)	4.7%	2.4%	No data	3%	2.4%

global emissions (Table 4-1) Augment these tonnages with those from facilities in societies striving to increase their industrial bases (e.g., China) and a projected global impact is truly frightening.

The "wake-up" call has been harkened to by many nations. In 1974 the United States government mandated manufacture of automobiles that use unleaded gasoline. This tact has been followed in most nations and has cut atmospheric lead pollution by more than half and the amount continues to decrease. The installation of efficient and effective emission controls (chemical scrubbers and electrical precipitators) and amount of emission permitted for industrial facilities has likewise been mandated by law. If legal regulations are not met, industries can be shut down or large and increasing "polluter pays" fines are levied. The "polluter pays" legislation is not entirely effective if a polluter deigns to pay fines and regulations are not enforced because of economic, political and social regulatory decisions or because of corruption. Thus, existing laws may not lessen the problem of emissions of potentially toxic metals to an atmosphere and the harm they cause to living populations over time through respiration. Similarly, they may not relieve atmospheric deposition of metals to waters and soils from which they can enter a food chain. Likewise, they may not relieve the threat of access to a food chain from the deposition of metal-bearing particulates onto soils. The purchase of pollution credits from one industrial pollution-generating facility by another, although legal, is useless because it does not lessen total potentially toxic metals emissions output and subsequent deposition.

Respiration of metal pollutants through dust released to industrial or mineral processing workplace atmospheres has had a significant impact on the health of workers. The most recognized health problems are "black lung" disease, silicosis, and radiation sickness in miners, and lung cancer in asbestos workers. However, many others have been documented (Table 4-2). This pathway to human health risks has been addressed in many, but not all societies.

Table 4-2. Potentially toxic metals in the workplace and health effects (adapted from text in Watterson, 1998)

Carcinogens which may occur in the workplace

Human carcinogens: As, Cd, Cr
Possible human carcinogens: Co, Pb, Ni
Lung carcinogens include: As, Be, Cd, Cr, Ni

Some reported cancers caused by or associated with certain occupations and industries

As: Lung, skin: Pesticides, other
Cr: Lung: Metals, welders
Ni: Lung, nose: Metals, smelters, engineering

Substances linked to occupations liver disease

As: Cirrhosis, angiosarcoma, hepatocellular carcinoma: Pesticides, wood, vinters, smelters
Be: Granulomatous disease: Ceramics

Substances reported to have damaged the kidney in the workplace

Cd: Nephrotoxicity: Welding, engineering
Pb: Nephrotoxicity: Chemicals, paint, batteries
Hg (inorganic):   Nephrotoxicity: Chemicals, paints

Table 4-2 (continued)

Possible factors influencing reproductivity outcomes based on experimental data

As: Fetotoxic, teratogen, transplacental carcinogen: Agriculture, wood preserving

Cd: Spontaneous abortions, impaired implantation, teratogen. Male and female damage: Engineering, chemicals, batteries, paints, smelting, mining

Cr: Teratogen: Chemicals, engineering

Pb: Decreased fertility, fetotoxic, impaired, implantation, teratogen, sperm damage, hormonal alterations: Various

Mn: Decreased fertility, impaired implantation: Various

Hg: Fetotoxic, teratogen, menstrual disorders: Chemicals, pesticides

Se: Fetotoxic

Tl: Fetotoxic

Triethyl Pb: Spontaneous abortions

Substances found in the workplace reported to have caused neurological damage

As: Peripheral neuropathy: Metal production, pesticides

Pb: Encephalopathy and peripheral neuropathy: General

Mn: Encephalopathy, ataxia, later Parkinson disease-like symptoms occur, acute psychosis: Engineering, aircraft industry, steel, aluminum, magnesium and cast iron production

Hg: Tremor, weakness, peripheral neuropathy is uncommon, chronic exposure leads to ataxia, mental impairment: Chemicals, pharmaceuticals, dentistry, plastics, paper, various

Ni: Headache: engineering

Tl: Encephalopathy, ataxia (high doses)

Sn (organic):   Encephalopathy

Substances known or associated with visual damage in the workplace

Pb: Optical neuropathy: Foundry industry

Hg: Cranial nerve palsies: Chemicals

Substances reported to have caused immune system effects in the workplace

Ni: Hypersensitivity: Metals engineering

Reported respiratory effects of certain workplace substances

Metals: Especially Pt, Ni, Cr, Co and V: Occupational asthma: Metal and engineering workers

Table 4-2 (continued)

Substances know to be absorbed through or damaging to the skin
in the workplace

As: Skin cancer: Agriculture, lead workers, dyers, copper smelters, brass
    makers, chemicals, textiles, painters, pesticide users
Cr: Allergic contact dermatitis: Metals and engineering workers
    Metals (many): Percutaneous absorption: Engineering and chemicals
Ni: Allergic contact dermatitis: Metals, engineering

Substances linked with cardiovascular toxicity

Arsine: Cardiac arrhythmia, peripheral
As: Myocardial injury
Sb: Hypertension
Cd: Hypertension
Co: Myocardial injury
Pb: Myocardial injury, hypertension

Reported adverse effects of chemical on the blood

As: Aplastic anaemia: Glass, paints, enamels, pesticides, tanning agents
Cu: Red blood cells: Engineering
Pb: Red blood cells, porphyria: General

Water

Another direct course to ecosystem inhabitants is through drinking
water which can carry a natural load of metals at pollutant concentrations (e.g., from weathering of undeveloped mineralization). Pollutant loading can also develop as metals are mobilized from natural
systems stimulated inadvertently by human activity. Metals' natural
contents in drinking water may also be augmented by anthropogenic
input originating from identifiable sources such as (toxic materials)
waste disposal sites or industrial/sewage effluent outfalls. Where
polluted waters are treated properly at collection facilities and then

distributed, potable water is available to human populations. However, one-third or more of the human population does not have access to clean water for drinking, cooking and personal hygiene, or water for sanitation. This is a health threat especially to babies and young children.

Food

The third direct route into the food web is via foods that have natural high contents of potentially toxic metals or that have bioaccumulated them in a growth environment. Plant uptake is one of the main pathways through which metals enter a food chain. This is from virgin soils, soils amended with heavy metals-bearing sludge (e.g., Cd, Cu, Pb and Zn), soils that incorporate pollutants from atmospheric deposition and those irrigated with metals-contaminated waters. The soils can be sources of polluted food crops such as vegetables and animal forage that bioaccumulate heavy metals. This pathway transfers the metals through higher trophic levels to humans.

Vardaki and Kelepertsis (1999) found an area of natural pollution of vegetation in Central Euboea, an island northeast of Athens with ultrabasic rocks and nickeliferous mineralization. They analysed soils, waters and plants for several metals (Ni, Cr, Co, Fe, Mn, Cd, Pb, Zn and Cu + As in waters). Soils had contaminant levels of Ni, Cr, Fe and Co with Ni and Cr contents higher than phytotoxic levels (100 and 75–100 ppm, respectively) compared to soil mean values of 2160 ppm for Ni and 1040 ppm for Cr. These soils may produce apparently normal crops that can be unhealthy for human and animal consumption because of the bioavailability of metals taken up by plants (a function of soil conditions). Element distribution and concentration in soils (and hence in plants) is controlled by geology such as the ultrabasic rocks and nickeliferous mineralization from which the soils in Central Euboea formed. Of sixteen water samples analysed seven had much higher than EU permissible concentrations of Cr, and As was higher in two. Of the vegetables growing in 25 soils with significantly high concentrations of Ni, Cr, Fe and Co, 19 were in the Ni toxicity range, 5 were in the Cr toxicity range with 12 above normal, and 3 were in the Co toxicity range with the rest above normal. Twenty-three of the samples were deficient in Cu and 7 were deficient in Zn whereas Pb was in the normal range for all samples. Vardaki and Kelepertsis (1999) suggested that studies be done to find which cultivars are most tolerant to heavy metal contaminants in the area and which should stop being cultivated to prevent human and animal toxicity. In addition, they propose that studies be made to determine if the present population has suffered health problems that could be correlated with the ingestion of pollutant concentrations of Ni, Cr and Co from food and water.

Similarly, aquatic life forms that bioaccumulate heavy metals from polluted waters and from contaminated sediments in a water column, or bottom sediments, can pass pollutants up the food chain. This can harm foodfish, predator mammals, and ultimatly a human population. Contamination of a growth environment from the atmosphere or via irrigation can be natural as described above but is most often the result of human activities.

Bioavailability, Bioaccumulation, Biomagnification

When a potentially toxic metal is mobilized from a source into an ecosystem, its chemical form will determine its bioavailability to a food web. In a review of metal accumulation pathways for marine invertebrates, Wang and Fisher (1999) found that both food and water can dominate metal accumulations depending on the species, the metal and food sources. They concluded that trace elements primarily in sea water as anionic species (e.g., As and Se) are mainly accumulated from food whereas uptake from water is important for metals associated with protein. However, bottom feeders such as polychetes obtain all of their metals burden from the ingestion of sediments because of high ingestion rates and low uptake from solution. Further, Wang and Fisher (1999) reported that inorganic and organic suspended particles critically affect exposure pathways of contaminant metals for marine invertebrates.

If a heavy metal does enter a food web, organisms can react to its bioavailability in different ways. Living things may discriminate against the uptake of one or more potentially toxic metals. Others can incorporate the metal(s) in their soft or hard parts in proportion to the concentration(s) in the growth environment and will not absorb excess contents but rather excrete/shed them. Still others are tolerant of heavy metals and will accumulate concentrations greatly in excess of amounts in a growth environment but without any damage to the host. These have been called accumulator or hyperaccumulator species. In vegetation from natural environments (soil free of anthropogenic input), for example, Brooks (1983) defined as accumulator species those plants with between 100 and 1000 ppm (dry weight) of a metal. Hyperaccumulator plant species are those with more than 1000 ppm. Feeders of the accumulator plants may further concentrate potentially toxic metals. In research on the tri-trophic transfer of Zn from soils amended with metals-bearing sewage sludge, Winder et al. (1999) measured a Zn transference and biomagnification along an exposure pathway from soil to plant to herbivore to predator. The concentrations measured were from soil (31.1 µg/g dry weight) to wheat (31.7 µg/g dw), to grain aphids (116.0 µg/g dw), to a predatory carabid beetle (242.2 µg/g dw). The amount of Zn in the predatory beetle was related to the amount of aphids eaten and the sewage sludge application rate (0, 10 and 15 tons/ha of dry solids). Swordfish might be considered marine hyperaccumulators of Hg. Samples of swordfish from a Smithsonian Institution laboratory collected many years ago contained about 500 ppb of the toxin. The maximum permissible concentration for Hg in fish was set at 500 ppb by WHO and national health ministries. Concentrations greater than 500 ppb analysed in fish from Lake Ontario during 1971/2 led to a temporary ban on the fishcatch. The Hg was attributed input from industries in Canada and the United States which discharged waste effluents into the lake. Regular consumption of fish with 500 ppb Hg, albeit a natural content, is considered a significant health risk and is discouraged.

Symptoms of Impact on Health of Ecosystem Life

As potentially toxic metals move along the food web from grazer to scavenger, from prey to predator, from vegetation to food animals to humans, bioaccumulation may be interrupted. It may also show bio-magnification up the food chain. Both bioaccumulation and bio-magnification can result in the attainment of toxic levels for a life form somewhere along the web which ultimately harm it or organisms at higher trophic levels in the food web. Different organisms or their physiological parts malfunction when they accumulate a certain level of toxins. This will often be masked by subclinical symptoms that are not easily detected. However, with continued exposure and additional accumulation, clinical symptoms of toxicity become evident. Age and body mass of a species influence the impacts of metal toxins which are often more damaging to younger groups in a population. Depending on the stage of toxicity in an organism, available therapies may or may not arrest a disease. Damage may be permanent and disabling. Metal toxicity will affect productivity and reproduction and bodes badly for the vitality of a species in an ecosystem. In some cases, however, toxi-city from the bioaccumulation and biomagnification of heavy metals or from deficiencies where these metals are essential micronutrients, may be reversible.

Antagonistic-Synergistic Effects

In other cases the impact of ingestion of pairs of metals to toxic levels may be intensified if the metals act synergistically to defeat normal body functions. Pairs of ingested metal toxins may also act antagonistically so that each cancels the negative effects of the other. Arsenic and Se are such a pair. On the down side for humans, however, if there is too much As, it interfers with the role of Se in metabolism. Antagonistic elements in an organism can also have a harmful effect if one metal inhibits the absorption of another that is an essential micronutrient. With time, this can lead to a deficiency of the micronutrient. For example, too much Mo in dairy cattle forage prevents absorption of the essential micronutrient Cu from the forage and hypocuprosis develops. The high Mo contents develop naturally from forage growing in soil formed from a rock (e.g., a black shale) with high Mo concentrations. Some consequencess of the Cu deficiency include a loss in weight gain, less milk production, and a drop in the reproduction rate (Webb, 1971). The effects of hypocuprosis are reversible by Cu solution injection and a change to forage with significantly lesser contents of Mo.

Conversely, prevention of absorption of a toxic metal can be healthy for ecosystem populations. Chaney et al. (1997, 2000) reported that when there is a common Cd/Zn ratio of 0.005/0.02 in contaminated soils (higher than in Zn-Pb ores), only rice and tobacco crops allow Cd to be mobilized from the soil in amounts that can be harmful to humans over a period of time. However, crops such as wheat and vegetables grown in aerobic soils may exclude Cd because the Zn inhibits uptake from soils and translocation of root Cd to edible tissue. For most animals, Zn inhibits Cd uptake and intestinal absorption thereby protecting them from Cd poisoning under most environmental conditions. Many plants suffer phytotoxicity from common heavy metals such as Zn, Cu, Ni and Mn at concentrations in plant shoots that are not harmful to livestock or humans by this common exposure pathway. There is ongoing research to develop transgenic plants with increased metal binding capacity to keep Cd in plant roots thereby reducing Cd accessibility to rice or tobacco crops (Yeargan et al., 1992).

CHAPTER 5

Contaminant/Natural Background Values: Timing and Processes

Natural geochemical background concentrations have been taken in the past as average crustal contents (Table 5-1). This is not compatible with environmental geochemistry research on specific pollution problems because of the great variation in composition of the rock types comprising the crustal surface (Table 2-1). This leads to variations in soil and other overburden chemistry. Chemical variations among flora and fauna likely reflect their growing and feeding in an area of varying geology. Added to this are the chemical changes rock materials undergo during weathering and from diagenesis. Natural background varies with other factors such as the sample used (Table 6-1), the size fraction or organism part analysed, and the analytical methodology employed.

Although absolute natural baseline contents may differ significantly from one sample type to another from the same region, geochemical maps prepared from them can be very useful in highlighting high value/low value areas. This was the case in northern Sweden reported by Selinus et al. (1996) for biogeochemical samples at two trophic levels: aquatic mosses and aquatic plant roots at a low level and moose tissue (liver and kidney) at a high one. Both trophic levels demonstrated the bioavailability of heavy metals in a region with an elevated Cd burden and the different trophic level geochemical maps were very similar. Salminen and Tarvainen (1997) emphasized that geochemical baseline values have regional and local variations that may have a wide

Table 5-1. Upper continental crust abundances of potentially toxic metals evaluated in this text with associated components common in rock-forming minerals (McLennan and Taylor, 1999; Taylor and McLennan, 1995). Values are in ppm unless otherwise noted

Upper Continental
Crust Abundances

Ag	50 ppb	Al	8.04 (wt%)
As	1.5	Ca	3.00 (wt%)
Be	3.0	Co	10
Cd	98 ppb	K	2.80 (wt%)
Cr	35	Mg	1.33 (wt%)
Cu	25	Mn	600
Fe	3.50 (wt%)	Na	2.89 (wt%)
Hg	80 ppb	P	700
Mo	1.5	Sr	350
Ni	20	U	2.8
Pb	20		
Sb	0.2		
Se	50		
Sn	1.5		
Ti	0.3 (wt%)		
Tl	750 ppb		
V	60		
Zn	71		

range according to changes in geology and type/genesis of overburden (soil, till in formerly glaciated zones). This will be the case whatever sample or analytical method is used. Such changes must be weighted carefully to make sound political decisions on environmental issues and in drafting workable environmental legislation.

As described in Chapter 2, weathering is the disintegration and decomposition of rocks by physical, chemical and biological interactive processes that can result in the formation of soils. During weathering, some chemical elements go into solution and move into fluvial and groundwater systems. Others are solubilized and recombine in a developing soil to form new minerals. Still others comprise minerals that are resistant to chemical decomposition and remain *in situ* as soil components. Resistant minerals, together with other soil minerals and organic components are subject to erosion by water, wind or ice and transport to depositional environments. The weathering leaves geochemical signals in soils and in eroded products that are tied to the original rock composition. Vegetation may take up chemical elements in proportion to their soil contents, but can also discriminate against uptake of proportional amounts, or bioaccumulate them in amounts that exceed soil concentrations. Depending on the vegetation involved, specific metals that move through plants into a food web may or may not reflect natural soil concentrations. Similarly, fauna feeding on vegetation may retain soil chemical signals whether natural from the original rock or enhanced by soil-forming processes. Fauna may further advance the bioaccumulation process.

Bioaccumulation by flora and fauna including humans is often organ or site specific (e.g., in liver, kidneys, nervous system, fat, leaves, twigs, wood, roots) and may be determined by analyses of the same. However, an organism's toxicity status may be first diagnosed by analyses of hair, nails, urine, blood, and exudates before determining if organ analyses are necessary. It is important to know health regulation action levels for contaminant metals in organism parts in order to determine if their measured concentrations are a threat to life forms being evaluated.

Contact with crustal solids influences the composition of terrestrial waters that flow on them or through them (in aquifers). Weathering of rocks to form soils also affects water metal contents. Geochemical signals in soils and waters may be altered by various processes. Among these are physical influences such as precipitation and atmospheric deposition, temperature, drainage and topography, and accumulation of resistant heavy minerals. Chemical influences such as changes in pH, redox potential, adsorption, absorption and desorption by suspended matter, and biological factors such as microbial activity and metal uptake by aquatic plants can change chemical composition of soil and water. In ensemble, these factors affect bioavailability and bioaccumulation in the food web. Because of such active processes and their influences, average chemical contents of specific rock-types (e.g., shales) are not useful as a measure against which to establish contaminant concentrations for comparative environmental geochemistry research.

The chemistries of specific rocks in an area and derivatives from them (soils, sediments, waters) together with geographic changes in them regionally or locally provide the natural background values. Potential contamination can be evaluated confidently only in comparison with these natural values.

Statistical Evaluation of Baseline/Background

Statistics is used to differentiate between natural (background/baseline) and contaminant concentrations (of metals and metalloids) in the various media that comprise living environments. Classical statistics assumes a normal or Gaussian distribution of measured values in a population (Figure 5-1). In this case, the (arithmetic) mean, X, can be

Fig. 5-1. Normal (Gaussian) distribution showing the graphical positions of some statistical parameters (after Krumbein and Graybill, 1965)

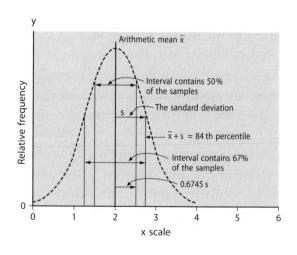

Variable: HG-PPB							
Bin	Lower	Upper	Count	Percent	Total	Percent	Histogram
1	75	198.125	10	9.3	10	9.3 :	*****
2	198.125	321.25	43	40.2	53	49.5 :	**********************
3	321.25	444.375	22	20.6	75	70.1 :	***********
4	444.375	567.5	11	10.3	86	80.4 :	******
5	567.5	690.625	3	2.8	89	83.2 :	**
6	690.625	813.75	7	6.5	96	89.7 :	****
7	813.75	936.875	2	1.9	98	91.6 :	*
8	936.875	1060	3	2.8	101	94.4 :	**
9	1060	1183.125	3	2.8	104	97.2 :	**
10	1183.125	1306.25	0	0.0	104	97.2 :	
11	1306.25	1429.375	0	0.0	104	97.2 :	
12	1429.375	1552.5	0	0.0	104	97.2 :	
13	1552.5	1675.625	0	0.0	104	97.2 :	
14	1675.625	1798.75	1	0.9	105	98.1 :	*
15	1798.75	1921.875	0	0.0	105	98.1 :	
16	1921.875	2045	2	1.9	107	100.0 :	*

Fig. 5-2. Frequency and distribution data for Hg in ppb for 107 marine sediments from the Voronin Trough, Kara Sea, European Arctic. The histogram representation illustrates the positive skew that is common in geochemical distributions (author's unpublished data)

considered the natural background values for a sample population in an area being studied. Values for the mean plus/minus one standard deviation $(X \pm 1\sigma)$ include 68.26% of the observations and are representative of regional fluctuations around the mean. Local fluctuations that can be established from values between $X \pm 1\sigma$ and $X \pm 2\sigma$ include an additional 27.18% of the measured parameters in a population. A total of 95.44% of the population values falls between the values $X \pm 2\sigma$ and likely encompasses both contaminant and natural contents. Concentrations for potentially toxic metals/metalloids $> X \pm 2\sigma$ are natural (e.g., from mineralization) and/or anthropogenic contaminants. These can be considered pollutant values if they degrade the normal functioning of living environments.

The parameters cited above to set background/contaminant values are not applicable to most metals/metalloids because few geochemical populations have normal distributions. Frequency plots are generally skewed with the positive skew most common for geochemical data (Figure 5-2). The distributions may approximate lognormality. Geochemists test for lognormality by preparing a cumulative

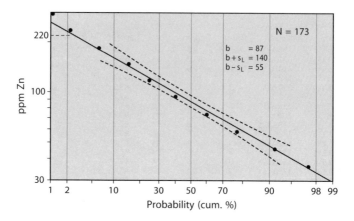

Fig. 5-3. Log probability plot of 173 soil Zn (B zone) values over a stockwork Cu-Mo zone, Tchentlo Lake, B.C. (after Sinclair, 1976). The near straight line plot indicates that the Zn has close to a lognormal distribution

frequency plot of the geochemical data on a logarithm/probability graph (Figure 5-3). A straight line plot indicates that a single lognormal population is represented. In this case, the geometric median (value at the 50th percentile) may be used to define a background or baseline concentration for the population. Values for regional and local background fluctuations, and high (and low) concentrations can also be determined at the appropriate cumulative frequency percentiles (e.g., at the 84th and 16th percentiles, and the 95th and 5th percentiles). In some publications values derived from lognormal/probability cumulative frequency plots are referred to as X_b, $X \pm 1b$, $X \pm 2b$ and $> X \pm 2b$. These concentrations likely differ markedly from those calculated assuming normal distributions but work well for the applied geochemist. Salminen and Tarvainen (1997) suggested that because most geochemical populations are skewed, the use of the median $\pm 1b$ best sets the range of natural baseline values.

Commonly, geochemical element distributions reveal contributions from more than one metal/metalloid population. In this case, low value (background/baseline) and high value (contaminant) popula-

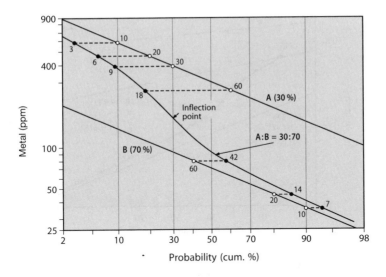

Fig. 5-4. An illustration of a log probability plot in which two sub-populations contribute to the total population (after Sinclair, 1976). The sigmoidal form of the cumulative frequency plot reflects the contributions from the sub-populations

tions can be extracted from cumulative frequency plots made using logarithmic/probability graphs. In geochemical studies, most plots show a sigmoidal curve when dealing with two contributing populations (Figure 5-4). The plots may present multi-inflections when dealing with more than two sub-populations (Figure 5-5). A median value and concentrations range for each sub-population can be extracted and baseline values established for each one.

An example of a simple graphical extraction of two sub-populations that contribute to a sigmoidal lognormal/probability cumulative frequency plot can be demonstrated using Figure 5-4 as follows: 1) the inflection point is at 30%; 2) this is taken to mean that class concentrations for the high value (contaminant) population represent 30% of their extracted values and those for the low value population (background) represent 70% of their extracted values; 3) when at least three extractions are plotted for each sub-population, the points are joined by a straight line which represents the statistically extracted

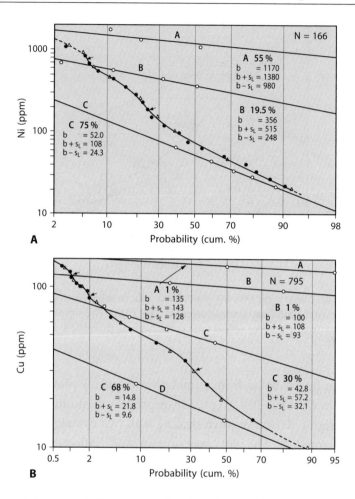

Fig. 5-5. A. Log probability plot of 166 soil (B-horizon) Ni values from an area near Hope, B.C. that includes a small Cu-Ni prospect (after Sinclair, 1976). Three sub-populations contribute to the cumulative frequency graph. B. Log probability plot of a population of 795 soil Cu analyses from an area near Smithers, B.C. (after Sinclair, 1976). Four sub-populations contribute to the total population

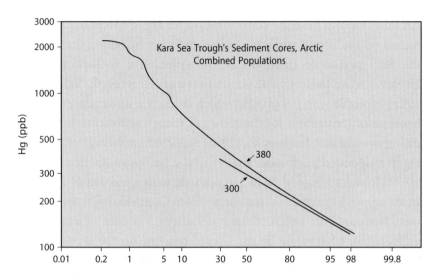

Fig. 5-6. Log probability graph of the Hg cumulative frequency in more than 400 samples from Kara Sea troughs, European Arctic. A baseline value of 380 ppb is determined by taking the median value of the total population. If the low value population is extracted, a more representative concentration of 300 ppb can be set as the baseline value

concentration range for each population. As seen in Figure 5-4, the high concentrations in the low value population can overlap with the low concentrations in the high value population. Stanley (1988) developed software that can extract up to five sub-populations from a dataset.

Rock type and geological-geochemical processes can change markedly in a relatively small area. This can affect sample chemistry. Overlapping populations extracted from a dataset allow better evaluations of whether geochemical measurements represent natural changes in background contents alone or anthropogenic input as well.

An example of the results of an extraction procedure is illustrated in Figure 5-6. This figure represents a lognormal-cumulative frequency plot of Hg values from a population of more than 400 samples from 52 cores collected in marine troughs from the Kara Sea, European

Arctic. A natural baseline concentration for Hg of 380 ppb was determined on the mixed (unextracted) population by taking the value at the 50th percentile (median) of the cumulative frequency graph. However, when the low value population was extracted, the value at the 50th percentile was 300 ppb Hg which is likely a more representative baseline concentration for the Kara Sea trough sediments. It is interesting to note that Joiris et al., (1995) reported an average of 300 ppb Hg in suspended particulate matter of the European Arctic seas. Skei (1978) found a range of 90 to 350 ppm Hg with a median of 250 ppm in the upper 5 cm of dried sediments from Gunnekleivfjorden, southwest Norway. This extraordinary pollution, 3 orders of magnitude greater than what might be considered baseline for the arctic region, was caused by discharge of inorganic Hg from a chlor-alkali plant during a 25 year period. The 250 ppm Hg is more than twice the median value reported for sediments from Minamata Bay, Japan. Forty-three villagers living along the bay died, 116 were afflicted with irreversible nervous system damage, and thousands of others were put at risk through the food web from the consumption over time of methyl Hg contaminated fish, a daily dietary staple.

Trend surface analysis is a method that is useful in following changes in natural background contents with changing geology and geological/geochemical processes in relatively large areas. The trend surface map contains numerical data and can separate variations in the geochemical data into systemic, regional and local components and residuals. Residuals are local geochemical values not related to the principal element distribution. Anomalous positive residuals reveal areas with potential for excess concentrations of chemical elements in an ecosystem and negative residuals highlight areas with potential for deficiency concentrations. The high value anomalies may represent natural contamination or areas of pollutant input from human activities.

Davis (1986) defines a trend "as a linear function of the geographic coordinates of a set of observations so constructed that the squared deviations from the trend are minimized". This is the surface of best fit using two independent geographic coordinates such as latitude and longitude. It can be represented by the equation:

$$X = a_0 + a_1 Y_1 + a_2 Y_2$$

in which a geochemical measurement X is a linear function of a constant value a_0 related to the mean of the observations plus a longitudinal a_1 coordinate and a latitudinal a_2 coordinate. In environmental geochemistry, a good prediction of the likely form(s) of a contaminant surface or distribution can often be made. Thus, by using a linear trend surface or a polynomial expansion that conforms to a projected non-linear surface (e.g., parabolic, saddle), it is possible to test for a surface of best fit where background values change in space (Figure 5-7A). The surface can bring out locations of samples or zones with anomalous or residual values. Figure 5–7B illustrates how positive and negative residuals appear in two dimensions and how the best fit for the dataset is the non-linear third order saddle surface. The most effective trend surface analysis can be prepared when there is a rather uniform distribution of sample sites and a spacing between them that is "tight" enough to be able to "catch" the smallest area of anomalous samples or zones that are expected to be found in an investigation. If this is not the case, computer interpolation to areas with no data diminishes the significance of environmental interpretation that can be made. Swan and Sandilands (1995) present an excellent discussion of trend surface analysis.

Multivariate analyses (e.g., cluster analysis or principal components factor analysis) of a dataset used to prepare potentially toxic metals pollutant maps can be useful in determining transport and depositional modes. An understanding of transport and depositional processes in an area can help delimit provenance areas and anthropogenic sources there that load pollutant metals into it.

For example, cluster analysis reveals significant associations among measured variables that can result in a fuller understanding of the physical, biological and geochemical processes active in determining

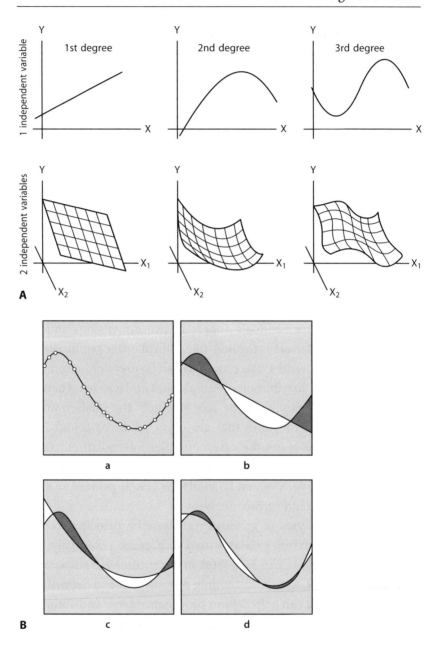

A

B

the chemistry of a sample suite. The analysis is designed to perform classification by assigning observations to groups so that each group is more or less homogeneous and distinct from other groups. In environmental geochemistry studies, cluster analysis is used to place elements, minerals, and size data, for example, into more or less homogeneous groups (communality) so that relations between groups is revealed. Samples can likewise be grouped by communality. Hierarchical clustering joins the most similar observations, then successively connects the next most similar observations to these (Davis, 1986).

In an investigation of As pollution in marine sediments deposited in Kara Sea troughs in the European Arctic, cluster analysis was done on data from 52 cores representing 476 samples. Sediment deposited before the industrial revolution had baseline values of about 21 ppm As but sediments deposited from the middle 1930s after industrialization in Siberia and the Ural mountains contained up to 710 ppm As, with several values greater than 200 ppm As and many values greater than 100 ppm (Figure 5-8). Cluster analysis (Figure 5-9) revealed that As forms a strong statistical grouping with Fe and P. This suggested that As could have been discharged into the sea, probably as the arsenite anion, from the Ob and Yenisey rivers drainage basins which contain many important industrial complexes. Once in the marine oxidizing environment with a pH $\approx$ 8.1, the arsenite transformed to the arsenate anion which was adsorbed by Fe oxy/hydroxides in

Fig. 5-7. A. Two- and three-dimensional representations of linear, quadratic and cubic (first, second and third degree) trend surfaces. B. The relation of measured values and the line on which they lie (a) to projected surfaces: (b – linear; c – quadratic; d – cubic) to find a trend surface of best fit (d). The plot locates zones of high (dark shaded) and low (unshaded) value residuals that could signal excess or deficiency contents available for a food web. Modified and compiled from Davis (1986)

Fig. 5-8. Regional distribution of high As values in cores from Kara Sea troughs, Arctic Ocean (after Siegel et al., 2001b)

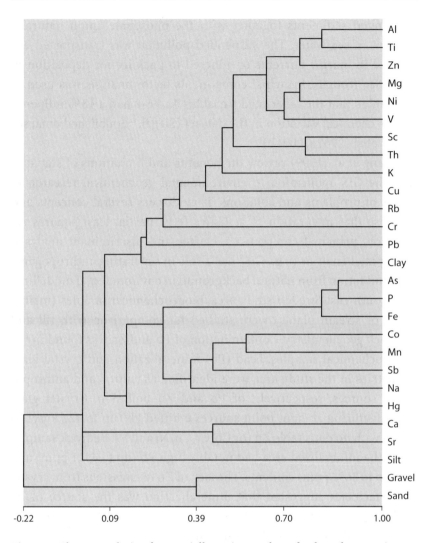

Fig. 5-9. Cluster analysis of potentially toxic metals and other elements in 476 samples from the cores represented in Figure 5-8

suspended sediments together with the phosphate anion naturally present in sea water. The suspended pollutant was transported entrained in marine currents or adhered to pack ice for deposition in Kara Sea troughs. Principal components factor analysis was used to determine that the associated variables As-Fe-P had a 18% influence on the chemical variation in the dataset (Siegel, unpublished data; see Siegel et al., 2001a, 2001b).

Zhang et al. (1999) review the benefits and limitations of statistics and the GIS application to environmental geochemical research in terms of problems and solutions. They discuss several concepts and propose that integration of statistics (e.g., partial least squares regression, principal components, cluster and discriminant analysis) and spatial analysis (e.g., GIS) has a role in separating anthropogenic contamination from natural background in environmental modelling. In a south-east Sweden study area, biogeochemical samples (mainly roots of stream plants) were studied for comparison with till and bedrock geochemistry. Contamination of Pb and Zn was found in the biogeochemical samples. Lead (Pb) mineralization and crystal glass industries in the study area were identified as natural and anthropogenic sources, respectively, of Pb and Zn pollution. Crystal glass manufacturing at eight point sources emitted Pb (up to 427 kg/yr in 1991) with no detectable Zn (or Co, Cr, Cu, Ni and V). Bedrock samples from the mineralized area had notably high Pb (up to 4311 ppm) and Zn (up to 3693 ppm) contents. The lack of Zn in emissions from crystal glass factories suggested that mineralization was the source of Pb and Zn contaminants in the stream plant roots. Statistical analysis supported these empirical conclusions on the origin of the Pb and Zn pollution and assisted the authors in distinguishing between natural and anthropogenic components.

Timing of Onset of Contamination

It is necessary to determine when contamination initially occurred in order to establish baseline concentrations for a site (e.g., in soil or sediment). When an onset timeline is set with an absolute age date, scientists can use pre-contamination values to calculate baseline values and enrichment factors (natural or anthropogenic). The known time factor also allows environmental scientists to quantify the rate of contaminant input, relate pulses of input to seasonal changes or to industrial operations (and reveal possible sources) and determine if and when potentially toxic metals in addition to those being investigated have become part of a contaminant inventory. This is a necessary first step before remediation can be realistically considered. The industrial revolution began in England more than 200 years ago and spread thereafter to other parts of Europe. This is considered by many as the onset of contamination. However, mining during Roman times and later caused damaging local metal pollution.

The onset of contamination can be determined for samples from very different environments. Sediments from marine, lagoonal, estuarine, lake, riverine and juxtaposed wetlands and floodplains may be dated using several methods. For example, absolute age dates using radiometric methods can be obtained for contaminated and baseline sections of cored samples using ^{210}Pb or ^{14}C radioactive decay systems. This depends on the time span being measured with respect to the half-lives of the isotopes. Average sedimentation rates are calculated from dated cores and extrapolated to non-dated sections to estimate a span of time represented by a core. Very recent sedimentation rates and ages of contaminated core sections can also be estimated from the first occurrence of the radioisotope ^{137}Cs which entered the earth's isotope inventory with hydrogen bomb testing (1953–1964). For example, the first occurrence of ^{137}Cs in cores from Manzala lagoon in the northeast Nile delta corresponds with the closure of the High Dam at Aswan during 1964 and yielded a sedimentation rate ≈ 0.5 cm/yr. This conforms with the rate calculated from ^{14}C dating on Nile delta cores. Availability of cheap electrical power stimulated major industrialization on the delta. There was poor regulation of industrial effluent and waste dumping into drains that emptied contaminants from as far away as Cairo into the southeastern part of the lagoon. The first occurrence of ^{137}Cs correlated well with the beginning of heavy metal contamination detected in the < 2 µm size fraction in lagoon cores (Figure 5-10).

The timing of contamination may also be established from trees by analysis of trunk borings for heavy metals. If high metal values are found along the boring, counts of tree rings can date when contaminants were introduced to the growth system. This has been an effective approach in areas downwind of smelters and electricity generating plants using coal and oil. Chimney emissions of metals as volatiles and particulates followed by their atmospheric precipitation results in the incorporation of metals in soils and subsequent translocation to trees, sometimes in anomalously high amounts. Baseline concentrations can be calculated from trees rings older than polluted segments.

Tree ring analysis has also been useful in following plumes of contaminated groundwater and tying the contamination to the development of waste disposal sites and the nature of the wastes they contain (Vroblesky and Yanosky, 1990). The premise is that the contamination is in groundwater tapped by tree root systems. Where the same tree species at about the same stage of maturity can be sampled along an aquifer flowpath, the rate and direction of plume movement can be

---

Fig. 5-10. Distributions of Hg, Pb, Zn and Cu in two cores from Manzalah lagoon, northeast Nile Delta (modified from Siegel et al., 1994). Core XV is close to the outfall of a drain carrying sewage and industrial wastes and Core XIV is about 2 km away along a current flowpath. A ^{137}Cs first occurrence marker relates well to the closure of the High Dam at Aswan, large scale industrial development and effluent discharge. Element concentrations drop significantly along the current flowpath

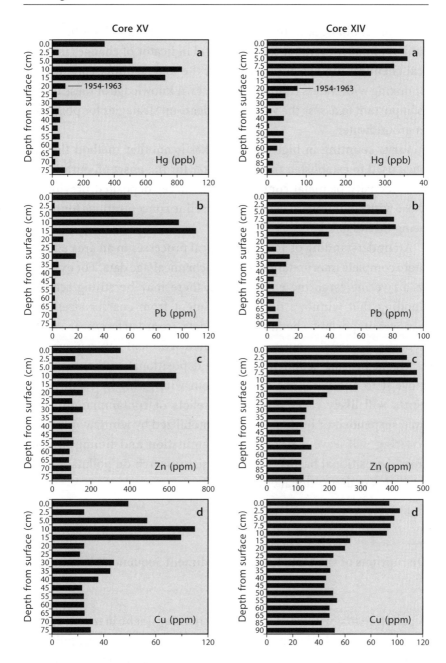

monitored. Changes in the concentrations of metals along the direction of plume movement can be a useful indicator of changes in physical-chemical-biological conditions of the water, aquifer composition or mixing with waters from other sources. A knowledge of these factors is important to assess the viability of clean-up strategies for pollutants in groundwater.

Varve counting in high latitude lakes is another method that has been used to establish a time parameter for the onset of, duration of, and intensity of metal contamination. Varves deposited previous to those with contaminant levels yield baseline concentrations for metals being investigated.

An understanding of active geological processes in an area allows a more complete interpretation(s) of geochemical/age data. For example, in a riverine-estuarine environment, there may be strong seasonal flooding that winnows fine-size sediment from coarser sizes. This removes important chemical signal data from surface/near-surface deposits. Seasonal scouring of deposited sediment can remove contaminant-bearing sediment from one depositional environment and move it to another. Core samples from winnowing or scouring regimes will likely carry little if any relics of contamination. However, suspended or bottom sediment mobilized by winnowing and/or scouring will leave a record of contamination and timing in cores from depositional basins. This can assist research on pollutant metal problems in depositional environments.

Disruptions of the Time Record in Sediment Sequences

A pollution history recorded in sediments may not be in true sequence because of bioturbation and/or solution, mobilization and precipitation of metals. Bioturbation causes a mixing of sediments from the surface to about 20–30 cm. In a core, this reworking of sediment dilutes high surface contaminant concentrations and results in strong pollutant values being displaced into near-surface samples (Figure 5-11). The mixing can also distort a time relationship based on absolute age-dating. If signs of bioturbation are not observed in cores visually and via radioagraphy, contaminant metals values and the time factor should be accurate if not affected by chemical processes.

The chemical processes of solution and mobilization along a geo-chemical gradient, and precipitation under new physical-chemical-biological environment conditions (collectively called diagenesis) can originate high values of some potentially toxic metals in surface/near-surface sediments. High values in core sections deposited during pre-industrialization times can also originate with diagenesis. Arsenic and Mn distributions in cores of fine-size marine sediments common-ly reflect the results of diagenesis. For example, in oxidizing marine environments (pH ≈ 8.1), As in the form of the arsenate species is adsorbed to a charged surface of Fe oxy/hydroxides. With burial and loss of oxidizing conditions because of organic decay and activity of aerobic bacteria, reducing conditions are reached and pH is lowered in the sedimentary sequence. At slightly reducing and acidic conditions, arsenate transforms to a mobile arsenite species. The arsenite moves up chemical gradient in the sedimentary sequence until it enters an oxidizing and higher pH environment where it precipitates as arsenate. Continuous activity of this type can build up natural con-taminant values for As in surface/near-surface sediment or in sedi-ment deposited during pre-industrialization time. A serious prob-lem in surface/near-surface post-industrialization sediment arises in distinguishing between the portion of an As high value that is natural and the anthropogenic part (Figure 5-11). This can be resolved to some degree by identifying a possible source or sources for As, estimating levels of contamination expected, and determining if there are viable pathways to depositional environments.

To determine if contamination does originate in a drainage basin that suffers strong seasonal flooding and loss of *in situ* contaminant signal by scouring in the fluvial system and coupled estuarine system

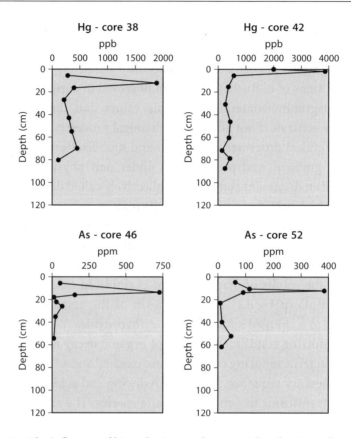

Fig. 5-11. The influence of bioturbation on downcore distributions of contaminant Hg and As in sediment cores from the St. Anna Trough, Kara Sea, Arctic Ocean (modified from Siegel et al., 2001b). Part of the As distribution is likely influenced by diagenesis

requires the use of process-related samples. These are overbank and floodplain sediments. Cores of flood-deposited samples contain chemical signals that reflect the history in time and space of potentially toxic metals in a drainage basin. Figure 5-12 gives the downcore distributions of Cu, Pb and Zn in dated overbank sediments from an area of the Harz Mountains, Germany. Anomalous values in one or more of the metals reflect ancient mining and smelting activities that were the sources of anthropogenic input to the river which deposited a con-

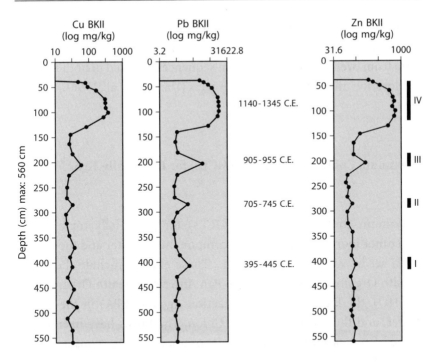

Fig. 5-12. Downcore distributions of Cu, Pb and Zn in a dated overbank sediment core from an area in the Harz Mountains, Germany. Four periods of mining/smelting are reflected by the sediment high value metal contents in the core. Natural baseline contents for the drainage basin can be calculated from sediments deeper than the oldest period of mining/smelting (modified from Matschullat et al., 1997)

taminant load in overbank and floodplain sediments. Mining and smelting had four periods of activity. Three of these were 40 to 50 years long between 395–445 C.E. (mainly for Pb), 705–745 C.E. (mainly for Pb) and 905–955 C.E. (for Cu, Pb and Zn). The fourth period, from 1140–1345 C.E. had greatly increased metal production for Pb, Zn and Cu likely from increased demand and improved mining and metallurgy methods. Core sections deeper than contaminated sections reveal baseline values for metals that are representative of rocks and soils in the drainage basin and natural physical-chemical-biological processes active there. Figure 5-12 also indicates that during non-active periods of mining and smelting, there was a relatively rapid return to baseline contents in the overbank sediments. Geochemical data from overbank and floodplain sequences are used in both mineral exploration and environmental projects (Darnley et al., 1996).

Maximum Contaminant Levels (MCLs) for Potentially Toxic Metals

Maximum contaminant levels (MCLs) for potentially toxic metals and other inorganic and organic components in water and foods have been set by various organizations. These entities include the World Health Organization (WHO), the Pan American Health Organization (PAHO), the Environmental Protection Agency (EPA) in the United States, and the European Health Union. The MCLs have been determined from careful and ongoing laboratory investigations and from medical records and observations on ingestion of potentially toxic metals, their bioaccumulation factors, and their impact on human health. The MCLs published by one organization do not always agree with those from others but are of the same order of magnitude and close in value. MCLs may change as new data on the relation between ingestion of excess or deficient amounts of metals and the health of inhabitants of an ecosystem are continually gathered by transnational groups, countries, provinces and states, over time and reviewed.

Table 5-2 gives a compilation of the MCLs established for potentially toxic metals in potable water. Data are also available from global and national organizations for allowable concentrations of heavy metals in the atmosphere, in agricultural soils, in foods and in the workplace. However, these are not as complete or as uniform as for drinking water. When a metal has concentrations that exceed its MCL in potable water or in published values for others media considered critical to human health status, alerted public health authorities work to put remediation action plans into operation.

Table 5-2. Maximum contaminant levels (also, action levels or maximum allowable concentrations) for potentially toxic metals in drinking water. Compiled from several sources such as WHO, EPA, EC, and others. Contents in air are given for some of the metals in concentrations as noted

	Water mg/l	Air
As	0.05	$0.01\ \mu g/m^3$
Be	0.004	$0.01\ \mu g/m^3$/month
Cd	0.003	$1 - 5\ ng/m^3$ rural
		$10 - 20\ ng/m^3$ urban
		cigarette 1 ppm
Co	1.0	$0.01\ mg/m^3$
Cr	0.05	$0.1\ mg/m^3$
Cu	1.0	$0.01\ \mu g/m^3$ rural
		$0.257\ \mu g/m^3$ urban
Fe	0.2	$6\ mg/m^3$
Hg	0.001	$10 - 20\ ng/m^3$
Mn	0.05	$0.05\ \mu g/m^3$
		$0.3\ \mu g/m^3$
Mo	0.04	$0.1 - 3.2\ \mu g/m^3$ rural
		$10 - 30\ \mu g/m^3$ urban
Ni	0.02	$0.002\ \mu g/m^3$ rural
		$0.015\ \mu g/m^3$ urban
Pb	0.0015	$1 - 2\ \mu g/m^3$
Sb	0.005	$0.2\text{-}2\ \mu g/m^3$
Sc	n. i.	n. i.
Se	0.01	$0.02 - 0.07\ \mu g/m^3$
Sn	n. i.	n. i.
Tl	0.002	$10\ \mu g/m^3$
Ti	n. i.	n. i.
V	n. i.	n. i.
Zn	5.0	n. i.

n. i. = not issued by health authorities.

CHAPTER 6

Indicator Media of Environmental Health Status

Critical phases in planning a study to assess the health status of an ecosystem are: 1) determining which samples can be used; 2) understanding what the sample represents in space (area and volume) and time; 3) knowing how chemical elements may be bound in a sample, physically and chemically; and, 4) establishing an ideal scale of sampling with realistic modifications as a function of where a sample type can be collected.

Many sample types are used to establish or monitor environmental geochemistry parameters used to study the health status of areas with existing problems or areas (Table 6-1). This includes using them to establish baseline concentrations against which the intensity of any negative impacts from development projects can be monitored. Monitoring has to be incorporated in programs for future industrial development. The samples are solids, liquids and gases, and life forms. Each is unique as to the environmental parameter(s) for which it yields data, their meaning, the area represented, the volume of earth material yielding the data, and normal (baseline) concentrations. Dated samples (e.g., from sediment cores) record changes, if any, in environmental parameters within the time represented by their ages. They allow an evaluation of physically and chemically accumulated elements as well as those that bioaccumulated in ecosystem organisms. In this text the focus is on heavy metals in the ecosystem.

Table 6-1. Selection of sample media that can be used in environmental geochemistry projects in which potentially toxic metals are the focus of research

A. Atmosphere:	emissions	
	volatiles, particulates, aerosols	
B. Hydrosphere:	precipitation (rain, snow, hail);	
	waters free of entrained solids from:	
	rivers and streams	
	aquifers and springs	
	lakes and ponds	
	estuaries and fjords	
	seas and oceans	
	wetlands	
	open taps	
	water treatment facilities	
	effluents	
C. Soils:	bulk	
	horizons A, B or C	
	separated size fractions from above	
	separated inorganic and organic phases	
	soil gases	
D. Sediments and rocks:	associated with water systems in the Earth's surface/near-surface environment:	
Suspended particulate matter:		
	inorganic, organic, colloid	
Bottom sediments:	surface, at depth in cores	
	separated size fraction from above	
	separated heavy mineral fraction	
	separated organic, inorganic phases	
Sedimentary, igneous and metamorphic rocks		
Ores		
E. Life forms:	flora:	
	trees, algae, plankton, other vegetation	
	morphological parts (e.g., leaves, twigs)	
	fauna:	
	shellfish and fish	
	birds	
	mammals (moose, seals, polar bears, humans)	
	separated fluids (blood, urine), hard (teeth, bone) or soft parts tissue (liver, kidney, muscle, hair, feathers)	
	insects	

Atmosphere

Discussions of atmospheric chemistry and the environment are often about increased CO_2 and its effect on global warming and sea level rise, and on SO_2 emissions and acid rain. Both originate from the burning of fossil fuels for transportation, in coal- and oil-fired power-generating facilities, for heating, and for industrial activities (e.g., smelting). However, during these activities, potentially toxic metals are also emitted to the atmosphere as gases, and as aerosols and particles. They are ingested by organisms during respiration in high concentration in proximate areas to a source and in lesser concentrations at more distant ones. Emitted metals are carried to ecosystems at the earth's surface via atmospheric deposition. They enter the food web from the air, mix with soils and sediments, and may create a health hazard for vital populations.

As noted in Chapter 2, there are natural contributions of metals to the atmosphere from volcanic eruptions, hot springs and wildfires. Loading from volcanic eruptions and wildfires comes in bursts and causes temporal distortions in atmospheric contents of potentially toxic elements that are then dispersed and diluted. Volatile elements such as Hg are emitted to the atmosphere naturally from soils, buried mineral deposits or subsurface rocks with high Hg contents, and by evasion from seawater. Atmospheric samples are collected for chemical analysis in proximity of and downwind of emission sources. Special equipment captures volatiles, aerosols and particulates, at intermediate heights above ground and at ground level where emissions issue forth at point sources.

By far the major anthropogenic loading of metals in the atmosphere is from coal- and oil-burning electrical power plants, from coal, oil and wood used in home heating, and from sulfide ore smelting operations. For example, Pirrone et al., (1996) estimated that 40% of Hg in the atmosphere over the former USSR is from coal combustion. The masses of metals emitted to the atmosphere from these sources are enormous. Pacyna (1995) reported that more than 20400 tons/yr of potentially toxic metals were emitted as air pollutants in the former USSR Arctic regions from industrial sources in the Urals, the Norilsk area (Siberia), the Kola Peninsula and the Pechora Basin (Table 2-5). These were from copper-nickel production, fossil fuel combustion, gasoline combustion, wood processing, steel and iron production, coal mining and phosphate fertilizer production.

Ingestion via respiration of a polluted atmosphere is a health hazard for human populations and can damage ecosystems thus putting many life forms at risk. The largest Ni-Cu smelting operations in the world are in Norilsk, Siberia, an industrial ciy of about 200,000 people. The original smelters supplied by Finland in 1934 were in operation during 1998 without the efficient chemical scrubbing systems and electrical precipitator systems that are in use in modern smelting operations. During 1995, the operators of the smelters reported an emission of about 2 million tons of potentially toxic materials, including heavy metals, to the atmosphere for the assessment of a "polluter pays" tax. On the basis of historical emission figures, the true amount emitted was likely much greater. As a result of the toxic emissions, human populations living near the smelters and downwind from them smelters suffer respiratory and other illnesses. The problem has been of such concern that the people of Norilsk are given a month's "vacation" annually to go to a spa where they spend time each day in respiratory cleansers. Areas around and downwind of Norilsk are "dead" from the outfall and precipitation of materials toxic to the living environment. The black area in Figure 6-1 shows the maximum degree of pollution where lichens have been totally eliminated. This important grazing ground for reindeer was destroyed. The area extends about 70 km to the SSE with the severe pollution zone adding another 50 km. The width of the maximum pollution zone is about 40 km with severe pollution adding another 20 km.

Potentially toxic metals from atmospheric deposition on a land surface are integrated with soil. They can become part of recharge to aquifers, or runoff into surface waters. Depending on element mobility

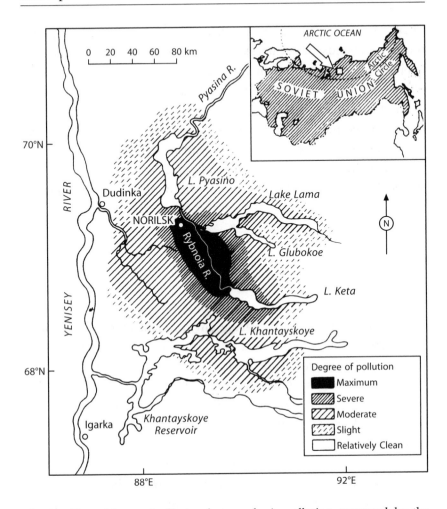

Fig. 6-1. The widespread effects of atmospheric pollution generated by the Norilsk industrial complex on lichens and other vegetation (after Klein and Vlasova, 1991)

in soils (a function of physical, chemical and biological influences – Chapter 3) and vegetation capacity to accumulate one or more metals or discriminate against their uptake, potentially toxic metals can access a food/water web. In addition, once in soils or associated waters the metals may be moved by erosion and in solution to fluvial systems.

They can be transported as particles (detritals) or soluble species through wetlands and estuaries (spawning grounds for food fish and habitats for many organisms). Here, some of the sediment load is filtered out and aquifer recharge may take place. Ultimately, the remaining load discharges into the marine environment. Toxic metals in the discharge may be immobile or may enter the marine food web where they can bioaccumulate in organisms and move up the food web to higher life forms possibly disrupting an ecosystem health status along the way. These scenarios will be described in following sections.

Soils

Soil provides the nutrient-bearing environment that sustains plant growth. Essential nutrient metals and other metals in food crops are translocated through soil into the food web. Natural contents of potentially toxic elements in soils are generally low unless soils develop from rock with high contents of one or more elements or from ore-bearing rock. Metal contents in soils may be greatly enhanced by human activities.

The genesis of soils was described in some detail in Chapter 2. As an environmental geochemistry sample, soil has to be considered in terms of the soil profile which develops over time as a temporal but ever changing end product of the interactions of physical, chemical and biological processes during weathering. Different environments characterized by rock type, climate (temperature and humidity/rainfall), vegetation and topography dictate the type of soil that forms over time, its chemical make-up and hence chemical mobility and bioavailability. Chemistry determines a soil's capacity to support different species of vegetation. This is in great part a function of the geochemistry of soil horizons (layers) and their dominant size/solid phase components. Table 6-2 provides a detailed description of the horizons

Table 6-2. Ideal soil profile (after USDA, 1951)

O: Organic horizons of mineral soils
     O_1: Formed or being formed in the upper part of mineral soils on top
          of the mineral part
     O_2: Dominated by fresh or partially decomposed organic matter
     O_3: Contains more than 30% organic matter if the mineral fraction con-
          tains more than 50% clay, or contains more than 20% organic matter
          if the mineral fraction does not contain clays; intermediate contents
          of clay require a proportional content of organic matter

A: Mineral horizons which consist of
     A_1: Horizons of accumulation of organic matter formed or being formed
          on or adjacent to the surface
     A_2: Horizons that have lost clay, Fe, or Al, with the resulting relative con-
          centration of quartz or other resistant minerals in the silt- or sand-
          size fraction
     A_3: Horizons dominated by A_1 or A_2 but transitional to an underlying B
          or C horizon

B: Dominant characteristic(s) as described in the following
     B_1: An illuvial concentration of clay-size silicates, Fe, Al, or humus, alone
          or in combination
     B_2: Residual concentration of sesquioxides or clay-size silicates, singly or
          mixed, that have formed by means other than solution and movement
          of carbonates or other more soluble minerals
     B_3: Thin layers of sesquioxides sufficient to impart conspicuous dark
          reddish colors, darker and more intense than those of the overlying
          and underlying horizons
     B_4: An alteration of material from its original condition that obliterates
          the structure of the original rock, results in the formation of clay-size
          silicates, free oxides, or both, and has a prismatic structure,
          blocky or granular if the textures are such that changes in volume
          accompany changes in humidity

C: Mineral layer or horizon excluding the parent rock that is similar or dis-
     similar to the material from which the solum has (presumably) formed,
     is only slight affected by the pedogenic processes, and has no properties
     characteristic of A or B

R: Underlying consolidated parent rock – granite, sandstone, limestone,
     others

that comprise an ideal soil profile. Because most potentially toxic metals in soils are bonded with greater or lesser strengths to clay minerals and sesquioxides, the B horizon provides a preferred sample for environmental assessments of a heavy metals threat to an eco-system through vegetation into a food web. A more accurate deter-mination of heavy metals mobility/bioavailability in the B-horizon can be made when the clay-size fraction is separated from a bulk/total soil (by sieving or pipette).

As stated in Chapter 2, the principal factors which influence an element's response to chemical mobilization during and post-weather-ing include the acid-base measure (pH), the reduction-oxidation potential, solubility, adsorption, temperature, time and concentration. These factors determine to a great degree an element's bioavailability in an ecosystem. Figure 3-2 illustrated how changes in environmental conditions (e.g., pH and redox potential) in a sedimentary environ-ment (e.g., soil) influences the mobility of several elements including some heavy metals. For example, As becomes more mobile as redox potential falls to lower reducing levels and pH becomes more acidic. The metals Pb and Hg also become more mobile under more acidic conditions but only when the redox potential rises to stronger oxi-dizing levels.

In the purest sense, the result of the weathering process is a soil that carries a chemical signature of minor and trace elements that com-prised the original rock. If the rock has mineralized veins that project towards the surface, they can be revealed by chemical analysis of a soil. Indeed, this is a basis for mineral exploration where mineralization is hidden beneath a vegetation-soil-rock cover. It is important to note that a sample of soil developed *in situ* represents only the area and volume of the collected sample. This may be as little as 25 to 50 cm^2 and 3000 to 6000 cm^3. Composite sampling can increase the area and volume represented by soil sampling. If a species of flora takes up one or more ore metals in proportion to soil concentrations, it may be useful as a biogeochemical indicator of subsurface mineralization. By virtue of the depth and lateral reach of root systems, vegetation samples will generally be more representative of a greater area and volume of soil than a correlative soil sample. This will be illustrated in the following section on life forms as samples in environmental geochemistry.

Some soils are designated as transported soils if the material on which a recent soil forms has been transported to a study area by an active geological process. Glaciers grind up earth materials and transport them in the direction of ice flow. When the glacier melts and retreats, it drops its solid load which is referred to as till. Soils form from these materials. At the base of this till-originated soil, a chemistry representative of the underlying rock can be found. However, in the upper part of the till soil, the chemistry can represent the geology of the area from which the till came. In many cases, glaciers transport earth material 3 to 5 km or more. If high values of metals are found in the upper till soil, they may reflect natural contamination and/or anthropogenic input from local or distant source areas up the ice flowpath. A knowledge of the flow direction can lead geochemists to possible source areas which can then be analysed for high values of metals found in the till soils.

Chemical analyses of soils and/or flora can reveal anomalous high metal values that alert authorities to the presence of pollutant elements in the ecosystem. These may originate from natural sources such as mineralization or a specific type of rock. For example, soil formed from the weathering of ultrabasic igneous rocks such as olivine basalts, pickrite basalts or gabbro pickrites inherit high concentrations of Cr, Ni and Co. These metals are commonly contained in minerals comprising the rocks such as olivine forsterite (Ni, Co), chrome diopside (Cr) and chromite (Cr). The metamorphic equivalent of these ultrabasic rocks that undergo mineral and chemical reorganisation and recrystallization is called serpentinite. Serpentinite forms under low grade regional metamorphism temperature and pressure abetted by active fluids. Serpentinites impart high contents of Cr, Ni and Co to soils that develop from it. Flora growing in these soils can take up Cr, Ni and Co in proportion to their natural baseline contents.

These baseline values will be significantly greater than in soils developed from the weathering of most other rock types. Flora may also accumulate these metals (> 100 to < 1000 ppm) or hyperaccumulate them (> 1000 ppm) to values much greater than soil natural baseline levels. Brooks (1983) described accumulator plants including bio-accumulators growing in New Zealand and New Caledonia in soils derived by the weathering of serpentinites. In either scenario, proportional or accumulator uptake by flora, metals can access the food web through vegetation. Anomalously high contents of potentially toxic metals in soils and their vegetation cover may also have sources in human activities. These would include metals in runoff from mines (acid mine drainage), in effluents from industries, in dumped wastes, in releases from buried wastes, and in atmospheric deposition of industrial emissions. High values of potentially toxic metals revealed by soil analysis alert environmental scientists to a possible threat to an ecosystem.

However, in many soils, the true threat to an ecosystem may be obscured by chemical analysis of a total soil sample, as already emphasized, metals in soils are mainly associated with the fine size fraction, especially the clay size (< 3.9 μm). The nutrient-bearing or principal growth soil horizon is dominated by this fraction which is comprised mostly by clay minerals and Fe-Mn oxy/hydroxides. Thus, chemical analysis of the fine-size soil matter, undiluted by coarser sizes, can better predict the potential of metals to enter an ecosystem and pollute a food web.

The ease with which a potentially toxic metal can access a food web through soil depends on how the metal is bound to a soil, the soil phase it is bound to, and its chemical form. Rulkens et al., (1995) characterized the physical states of pollutants in soils and sediments as being present as particulates, liquid films, adsorbed ions, absorbed ions, and liquid phases in pores (Figure 6-2). Kabata-Pendias (1993) identified five solid phases in soils from Poland from which metals were more or less able to enter the food web. These are from readily soluble phases, from exchangeable sites, from Fe and Mn

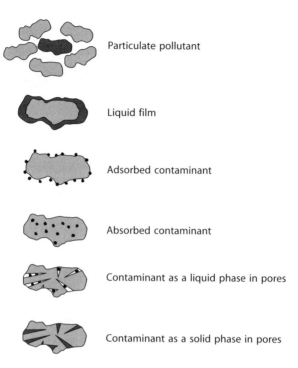

Particulate pollutant

Liquid film

Adsorbed contaminant

Absorbed contaminant

Contaminant as a liquid phase in pores

Contaminant as a solid phase in pores

Fig. 6-2. Physical states of contaminants in soils and sediments (after Rulkens et al., 1995). These affect potentially toxic metals chemical reactivity, mobility and bioavailability

oxy/hydroxides, and from organic matter. Metals in residual phases are not easily released to solution under most environmental conditions. Each of the seven metals studied by Kabata-Pendias (1993) is unique in the way it is bound (Figure 6-3). Chromium, for example, is mainly in the residual phase with a minor amount ($\sim 2\%$) in an easily soluble (readily available) phase. More than 40% of Cd is in a residual phase but > 20% is available from easily soluble and exchangeable phases.

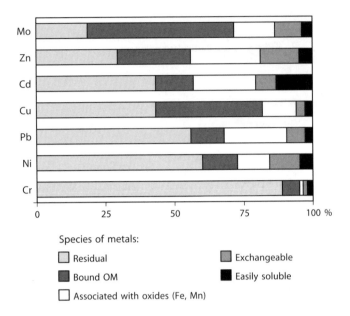

Fig. 6-3. Distribution of some potentially toxic metals in solid phases of soils in Poland (after Kabata-Pendias, 1993). Distributions in sediments are similar

Waters

Natural waters in streams and rivers, ponds and lakes, and aquifers and springs have a chemical element content that reflects the chemistries of rocks through or over which they flow. Thus, water chemistries can vary greatly in nature with rock compositions. In addition, during flow, factors such as temperature, pH, redox potential, adsorption, and bacterial activity can bring about changes in water chemistry. Natural waters contain entrained suspended particulate matter and colloids (nm to 1 μm) that can affect analytical results. This requires that immediately after collecting a water sample, entrained solids are separated by vacuum-filtering in the field through special filters that remove the > 0.45 μm size fraction. After filtering, water samples are acidified in the field. This yields a more accurate geochemical analysis for the water and at the same time provides suspended matter for water-sediment element partitioning research.

In spite of the filtration, water chemistry can be influenced by colloids that pass through the filter (< 0.45 μm). Degueldre et al. (2000) investigated this factor in different aquifers. Colloids identified were comprised of inorganic clay and oxide minerals, clay and oxide minerals coated with organic material, organic matter associated with inorganic phases, and organic matter. Their concentration and stability affect their capacity for attracting and transporting pollutants. Degueldre et al. (2000) found that colloid concentration was a function of pH, redox potential, contents of Na, K, Ca, Mg and dissolved C_{org}, and the physical-chemical steady state nature of a hydrogeochemical system. Colloid stability which also influences colloid concentrations in an aquifer was determined to be a function of attachment rate. Greater amounts of stable colloids in aquifer systems can move more pollutants, including potentially toxic metals, farther along a flowpath to pumpage zones where the pollutants can intereact with an ecosystem. The separation of colloids from aquifer waters to obtain a sufficient mass for chemical analysis is a time-consuming process but certainly is a promising field of research in studies of subsurface water transport of heavy metals.

The chemistry of riverine or aquifer samples is a cumulative representation of upflow water chemistries. It represents an extended area whether from surface drainage alone, inflow waters to streams and aquifers, and drainage or leakage from human activities. Thus, high values of contaminant metal in waters can be natural and represent a rock type change (Table 2-1) or the existence of mineralization. The high values can also originate with industrial activities (e. g., acid mine drainage, effluent from chemical, plastics, metal plating, wood preserving), or leachate from waste disposal sites. Careful tighter-spaced upstream sampling can delimit areas from which high metal concentrations originate. The closer water sample spacing is, the better a contaminant source (or sources) can be localized.

Lakes and ponds carry water chemistries of surrounding drainage areas and physical, chemical and biological processes active there as well as within the lacustrine environments. Source areas for metal contaminants in lakes and ponds may be revealed by a systematic sampling of inflow zones.

Waters from estuaries or fjords carry chemical input of end members from fluvial systems and the ocean. Estuarine water chemistry is also influenced by daily tides and major storms and storm surges inland. This results in waters with blended chemistries. Dilution by ocean water masks the introduction of potentially toxic metals but these may be revealed by analysis of associated fine-size sediments which sorb and concentrate the metals. As is the case with riverine and lake/pond environments, runoff from urban and agricultural areas into estuaries can strongly impact water chemistry. Ocean waters may have detectable metal contamination only where sampling is done proximate to discharge areas of polluted rivers or at effluent outfalls. Otherwise a chemical signal is diluted and diminished by both water mass and reactivity of metals with suspended and bottom sediment or sorption of metals by associated flora and fauna. These sample types are discussed with respect to heavy metals in the environment in the following chapter sections.

Sediment and Rock Associated with Water Systems

Soil formation and factors that affect soil chemistry and the distribution of potentially toxic metals was discussed in Chapter 2. Water plays its role in soil genesis and chemical element mobility during rock degradation, the release of resistant minerals and the reorganization of dissolved and solid decomposition products into new minerals. These soil minerals will suffer erosion and become part of the sediment cycle. They are transported mainly by fluvial waters (but also by glaciers and wind) to ponds and lakes, fjords and estuaries, seas and oceans. Sediment is deposited closer or farther from its input zone according to size and specific gravity, a sorting influenced significantly by tidal and current movement in large lakes, seas and oceans. The coarser sediments (gravel, pebbles and sand) deposit closer to shore and retain chemical signatures of rocks from which they disintegrated. Silts, especially the finer size, will have a chemistry similar to that expected from residual detritus but may adsorb metals as well.

Clay minerals and iron and manganese oxy/hydroxides (amorphous and crystalline) are the most reactive phases with respect to sorption of metal ions and anionic complexes in water environments because of their charged surfaces. The uptake is cumulative. Strong enrichment of metals by clay-size sediment and colloidal matter is documented in suspended sediment, in deposited sediment, and in their equivalent shale and iron and manganese oxide mineral and rock phases.

Timing of sampling is a factor to be considered in environmental studies in which seasonal dynamics can affect analystical results. Stecko and Bendell-Young (2000) described the partitioning of Cu, Pb and Zn among four extractable fractions in deposited sediment and suspended particulate matter collected monthly from August, 1994 to August, 1995 from the Fraser River Estuary, British Columbia, Canada. Copper, Pb and Zn were analysed in an easily reducible fraction (Mn oxides + amorphous Fe oxides), a reducible fraction (Mn + Fe oxides), an organic fraction (organically bound), and an aqua regia or "total" (residual) fraction using the sequential extraction sequence illustrated in Figure 6-4. There were large and significant differences in seasonal contents of Pb (Figure 6-5), Cu and Zn in the extracted fractions of suspended sediments. Stecko and Bendell-Young (2000) found that bioavailability of the metals studied and likely others in the suspended particulate matter have their greatest seasonal fluctuation and bioavailability from November through March. This is a low flow period that facilitates available metal uptake by filter-feeding organisms in the Fraser River Estuary food web. The deposited sediment samples revealed relatively little seasonal change. Certainly, the seasonality

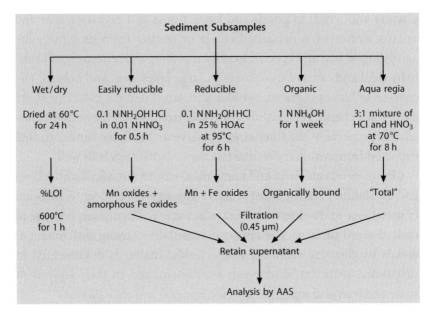

Fig. 6-4. Simultaneous extraction scheme used for determining metal partitioning in suspended and deposited sediments (after Bendell-Young et al., 1992)

influence on bioavailability of metals from samples such as suspended particulate matter must be taken into account in planning evaluations of potentially toxic metals access to a food web.

Sorbed or precipitated metals may be from the waters themselves but also may originate from within a sediment sequence where *in-situ* conditions can stimulate mobilization of some chemical elements. These elements move up a chemical gradient until changed conditions lead to adsorption or precipitation at or close to the sediment-water interface. Potentially toxic metals in the water column and as sorbed species may enter food chains and be harmful to organisms that bio-accumulate them in organs and tissue.

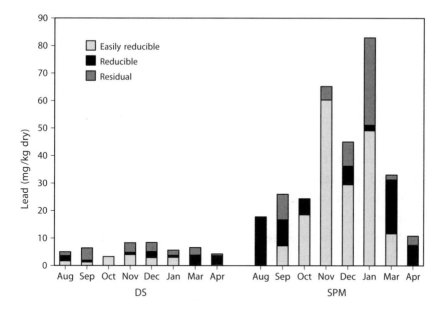

Fig. 6-5. Changes in seasonal extractability of Pb from three phases in deposited sediments and suspended sediments (after Stecko and Bendell-Young, 2000)

Life Forms

Organisms can be effective monitors for measuring the input and accumulated concentrations of potentially toxic metals directly in water, an atmosphere, and along a food chain. These are called biomonitors. Biomonitors are sensitive to changes in other ecosystem parameters (e.g., pH, redox potential, turbidity) that may render it an unsafe habitat.

Biomonitors are used to establish geographic distribution (when organism ranges are reasonably well known) as well as temporal variations in bioavailabilities of heavy metals. As such they have to be metal accumulators. They can provide environmental scientists with both a time-integrated measure of metal loadings and mass per area metal loading values that are of ecotoxicological relevance (Rainbow,

1995). Organism heavy metals uptake can be affected by species, by seasonal growth changes (often temperature influenced), by the time of exposure, and by the bioavailable species of a metal ion. In addition, the results of chemical analyses can be markedly different depending on the laboratory preparation of flora/fauna before analysis. Geo-chemical data from biomonitors can be best compared when the constancy principle is followed. The species, age or state of maturity, orientation (of fixed or attached forms) in the growth environment, and field and laboratory protocols are examples of factors which should be kept constant or nearly so during environmental geoche-mistry studies.

Many life forms have been and continue to be evaluated as monitors of potentially toxic metals in the atmosphere, and in waters, soils and sediments that comprise ecosystems on land, in estuaries, in seas and in oceans. These include terrestrial and aquatic vegetation, bacteria, molluscs (gastropods, bivalves), leeches, worms, insects (bees, mayfly larvae), fish, birds (seabirds), and mammals (e.g., reindeer, moose, seal, polar bear, humans). Some of these are considered in the fol-lowing paragraphs.

Flora

The flora as food crops, as forage for food animals, and as non-comes-tible vegetation may reveal the toxic metal(s) status in an environ-ment. Plants can be indicators of what is in the atmosphere, soils and waters that feed the growth medium. In a geobotanical sense indicator plants are those that are in equilibrium with respect to their ecosystem requirements such as nutrients, solar orientation, soil, pH, redox potential and humidity. The nutrients may include one or more metals. This means that where indicator plants with known metal requirements are found, there is enough of a concentrations of the metal(s) in the growth medium to supply the needs of the species. The absence of indicator plants means nothing about metals in the eco-system since there may not be enough of one or more metals to satisfy a plant's requirements. For example, the foodcrop corn (maize) and the coffee plant are sensitive to Zn deficiency and under this condition will not sustain a healthy growth.

The best indicator plants for service as biomonitors should incorporate a metal in proportion, or nearly so, to its concentration in soils and soil waters and have a sufficient geographic distribution to allow good density sampling. If a plant's chemistry is not directly related to its environment, the biogeochemical data can be misleading in a geochemical interpretation for several reasons. First, some vegetation can bioaccumulate metals from baseline concentrations in soils and signal higher values for the growth environment than are really there. Vetevier grass acts this way but as such lends itself to soil bioremediation projects. Similarly, water hyacinth incorporates high contents of potentially toxic metals in plant tissue by rhizofiltration and can be used in remediation and/or control of the metals in engineered, managed wetlands. In these cases, there must be an assessment of how the use of metal-laden vegetation could impact an environment and of protocols on how the vegetation can be safely disposed of. This is discussed in Chapter 8 on remediation. Second, some vegetation can discriminate against uptake of metals that could be harmful to them and thus not provide an accurate biogeochemical signal of metals in a growth environment. Third, besides the observation that some vegetation is able to selectively accumulate certain potentially toxic metals, vegetation can be selective in the organ which uptakes metal(s).

Thus, in some vegetation, metals may bioaccumulate more of a bio-available metal in leaves, twigs, bark, wood or other parts of vegetation. Or the accumulation may be greater in first year growth (e.g., needles, twigs) than in second year growths. One result of this is that comparisons for highlighting areas with high and low metal contents can be made only between like parts from plants at equivalent stages

Table 6-3. Arsenic contents of different organs of different ages from *Pseudotsuga menziesii* (douglas fir) growing at Canadian and Alaskan mining areas (after Warren et al., 1968)

Location	Sample	Organ	As (ppm)
H.B. Mine	64-3	First year twigs	510
		First year needles	120
		Second year twigs	70
		Second year needles	25
H.B. Mine	64-5	First year twigs	780
		First year needles	450
		Second year twigs	280
		Second year needles	60
King Vein of the Bralorne Mine	64-2	First year twigs	2110
		First year needles	1060
		Second year twigs	1390
		Second year needles	180

of maturity. Table 6-3 generated for biogeochemical mineral exploration illustrates the significant differences that can exist. Similarly, metals may accumulate more in edible portions of foodcrops than in non-edible parts. Table 6-4 gives examples of the relative accumulations of Cd and Pb in edible portions of foodcrops and contents of Cu, Ni and Zn in their leaves. For example, Cd and Pb have high contents in celery but low contents in potato and maize whereas Cu, Ni and Zn have high accumulations in sugar beet leaves but low accumulations in onion leaves. In addition, when striving to identify high value and background concentration areas, comparisons can be made (vis-a-vis the constancy principle) only between vegetation of approximately the same state of maturity (mass, age). Whole (small) plant analyses are of questionable use since high biogeochemical signals in one vegetation part may be diluted and masked by low concentrations in other parts.

Table 6-4. General high and low accumulations of Cd and Pb in edible portions of food crops and of Cu, Ni and Zn in non-edible leaves of food crops (after Alloway, 1995)

Metal	High accumulations	Low accumulations
Cd	Lettuce, spinach, celery, cabbage	Potato, maize, french beans, peas
Pb	Kale, ryegrass, celery	Some barley cvs, potato, maize
Cu	Sugar beet, certain barley cvs	Leek, cabbage, onioin
Ni	Sugar beet, ryegrass, mangold, turnip	Maize, leek, barley cvs, onion
Zn	Sugar beet, mangold, spinach, beetroot	Potato, leek, tomato, onion

In planning an environmental assessment of heavy metals in an ecosystem, indicator vegetation must be identified versus accumulator or discriminator vegetation. Many of these have been catalogued and listed for regional and local projects by geochemists using biogeochemistry in mineral exploration programs (Brooks, 1983; Brooks et al., 1995; Kovalevskii, 1991). Vegetation can provide data from an area and volume of soil that is much greater than that of a soil sample (e.g., 15 × 15 × 25 cm volume) because root systems extend deeper and/or farther laterally and tap a greater volume of soil (Figure 6-6). Samples (e.g., leaves, twigs) from a shrub with a root system extending through a cubic meter contain chemicals from > 175 times the volume of a typical soil sample. Samples from a tree with roots that reach 20 cubic meters of soil have chemicals from > 3500 times the volume of a soil sample.

Perhaps the best example of biomonitors in the flora taxa are lichen and mosses. Their use as bioindicators of atmospheric pollution (natural and anthropogenic) is established. The factor that makes them ideal for atmospheric monitoring is that they absorb little if any

Fig. 6-6. Relative volume of subsurface represented by soil and plant samples (after Siegel and Segura P., 1995)

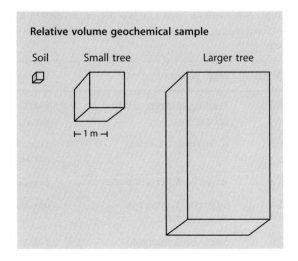

Relative volume geochemical sample

Soil   Small tree   Larger tree

⊢ 1 m ⊣

nutrient or other chemical elements from the underlying soils or rocks they carpet. Instead, they collect nutrients from the air and from atmospheric deposition either dissolved in rain water or as particulates. Potentially toxic metals from industrial emissions are likewise absorbed into the tissue of lichen and mosses. The different tolerances of lichen and mosses to pollutants are reflected in the distribution (abundance and health) of individual taxa. Their chemical composition can thus be used to ascertain the chemistry of industrial emissions and define the areas that have suffered vital damage or death around and downwind of emissions from industrial complexes as illustrated in Figure 6-1 for the area affected by smelter emissions at Norilsk, Siberia. Kansanen and Venetvaara (1991) studied the distribution of airborne Cr and Ni dust comparing mosses and lichen, pine bark, needle litter, earthworms and moths. They demonstrated that for Cr and Ni the mosses showed most effective accumulation of heavy metals with the lichen giving almost as good results. The relation to atmospheric deposition was clear and analytical replication was good. Kansanen and Venetvaara (1991) concluded that mosses and lichen were more effective in collecting heavy metals at low to moderate pollution loads rather than near the pollution source where their life functions declined or the vegetation died. Analytical data from mosses or lichen can reveal what atmospheric deposits contained during the life of the vegetation, and also the mass deposited per square meter of terrain. Analyses of mosses and lichen during increments of one or more years can be used to monitor atmospheric deposition with time (Selinus, 1995).

Fauna

Fauna are also useful as environmental monitoring samples of potentially toxic metals in an ecosystem. A wide array of faunal parts are studied with the aim of finding those that can be effective and reliable biomonitors. These include organs, other tissue, fluids and solid parts of humans, other mammals, fish and shellfish, foul, worms, and insects.

Shellfish and fish have been suggested as monitors for specific regions. For example, gastropods and bivalves from the coastal estuarine-deltaic regions of West Bengal, India were found to accumulate Hg in their shells (Bhattacharya and Sarkar, 1996). They were proposed as sensitive bioindicators of Hg pollution for this and similar aquatic environments where there could be Hg-bearing industrial discharge. In areas where there are significant annual temperature changes in bivalve habitats such as Lake Erie, Canada, some heavy metals in shells (e.g., Pb, Cd, Cu, V) show increased uptake with warmer water (Al-Aasm et al., 1998). This may increase interest in research on the use of shells as biomonitors. In addition to a temperature influence on shell chemistry, experimental studies by Powell and White (1990) on barnacles reported that Cd uptake was affected by salinity during a 50 day exposure period diminishing their potential to monitor Cd. These researchers suggested that barnacles may be suitable monitors of ambient Cu, Zn and possibly Pb. Field testing is clearly necessary. Metal pollution may not affect the health status of a lifeform directly but can affect the ability to survive in an environment. Lefcourt et al. (2000) observed that snails living in 10 polluted lakes downstream from a superfund cleanup site in the Coeur d'Alene basin of Northern Idaho failed to exhibit antipredatory behaviour when compared with snails from 14 unpolluted reference lakes. This could lead to declines in snail populations in polluted lakes and disruptions in food webs that would impact other organisms in an ecosystem. In a sense, this type of observation can alert scientists to pollution problems in aquatic environments.

In 1971, Warren et al. tested the use of the chemistry of lake trout livers for mineral exploration in lakes in British Columbia that could receive metal-bearing waters and sediments from surrounding areas. The lakes were far from industrialization and population centers. Trout livers from 30 lakes had Cu contents less than 60 ppm (wet weight) which was taken as a natural baseline. At the 17 lakes where trout liver Cu contents was greater than 60 ppm (wet weight), 4 reflected significant Cu mineralization, 7 were near known geological environments where detailed geochemical exploration for mineral deposits was planned and 6 were in areas where the geology was little known and were considered exploration targets. Environmental geochemists built on this concept and have investigated the use of fish and shellfish as biomonitors. They found that some lake- or lagoon-bound fish and shellfish, those restricted to riverine or estuarine biomes and to nearshore marine waters can reflect water chemistry in their soft parts. This makes them suitable to serve as biomonitors of natural chemical systems or those impacted by anthropogenic heavy metals input. Norrgren et al. (2000) evaluated the threespot tilapia as a bioassay form for areas of the Kafue River which receive effluent discharged from mining and other industrial point sources in the Zambian Copperbelt. Threespot tilapia caged for two weeks downstream of pollution sources rapidly bioaccumulated several heavy metals (e.g., Cd, Co, Cu, Cr and Ni). This is a bioindicator of metal contamination that may affect aquatic animal health. The metals uptake by threespot tilapia signals that mining and industrial waters should be treated before discharge in order to preserve sustainability of Kafue River ecosystems.

In 1996, Dietz et al. reported that a substantial portion of arctic marine mammals and seabirds have Cd and Hg levels that exceeded the Danish standard limit. Subsequently, Dietz et al. (2000) studied marine biota from Greenland and found that Cd, Hg and Se increased in concentration toward higher trophic levels. They also observed that concentrations in marine biota were higher than those found in freshwater and terrestrial ecosystems and attributed this to the presence of longer food chains. In a food web, predators that feed on fish and shellfish may bioaccumulate heavy metals and themselves pass the metal toxins up the chain and damage the health of organisms. Nendza et al. (1997) suggest that in order to assess hazards and risks to marine life such as Cd and Hg in seabirds and mammals, prey fish body burdens should be determined, not just the body burdens of the higher trophic levels.

Seabirds feed on fish. Their wide range and high trophic level make them useful as biomonitors of potentially toxic metals. If the prey fish for a seabird species is carrying heavy metals pollutants it follows that birds may reflect this for one or more than one of the metals in a specific part of the seabird. The seabirds fairy prions in Tasmania illustrate this relation. Cadmium contents of liver and muscle tissue from fairy prions in Tasmania were significantly higher closest to a potential source of pollution but contents of Cu, Zn and Pb were not. Thus the fairy prion could be useful in monitoring Cd pollution from the source to the marine environment to fish and then seabirds (Brothers and Brown, 1987). Monteiro et al. (1998) investigated Hg levels in feathers and food of 6 seabird species from the Azores archipelago. They found that feather and food mean concentrations were 4 × higher in seabirds feeding on mesopelagic prey (those below the zone of photosynthesis) than those feeding predominantly on epipelagic prey. Concentrations in feathers were 150 × those in food.

This represents the highest methylmercury biomagnification factor reported for consumers in aquatic food chains. Monteiro et al. (1998) concluded that seabird feathers have significant value as monitors of Hg risks in marine ecosystems. Savinova et al. (1997) studied six species of seabirds from seven different regions in the Barents Sea (n = 379). The species *Kittiwake* had significantly higher mean dry-weight levels of Cd (48 ppm), Zn (126 ppm) and Hg (1950 ppb) from the Ny-Alesund location than other locations. The highest As concentration (about 160 ppm against a mean of 95 ppm) was significantly higher by one to two orders of magnitude in Kittiwakes from Guba Chornaya Bay than at other sites. If the high As content originates with undersea atomic weapons testing as proposed by Loring et al. (1995) for a high sediment value of 305 ppm for this location, As contents of Kittiwake livers may be a bioindicator. If there is a pathway from crustacean to Kittiwake, crustacean foodweb prey and crustaceans should be evaluated as biomonitors.

Similar to seabirds which have well-defined ranges that help make them geographically important biomonitors, insects may also be effective biomonitors. Bromenshenk et al. (1983) determined that the chemical composition of honey bees or the pollen they carried (a floral element) could be used to monitor regional pollution in industrialized areas where the pollution originated from atmospheric deposition on vegetation and soils. They based their research on the knowledge that honey bees have an active range of about 3 km and the concept that nectar and pollen they carried to hives could reflect the chemistry of plants which in turn reflect the chemistry of soils and the rocks from which the soils formed. Bromershenk et al. (1983) also found that about 20 % of the variation in the chemical parameters was seasonal and depended on the distance from the source of contamination. Veleminsky et al. (1990) demonstrated that honeybees and their products were useful as biomonitors for Pb, Cd, Cu and Zn pollution in central and southern Bohemia. On the basis of these studies, it was proposed that trace metal concentrations in honey and pollen could be used as a reconnaissance sample in the search for mineralization.

Recently, Sjøbakk et al. (1997) evaluated the chemistry of three orders of fly larvae including mayfly as possible biomonitors of heavy metals in two rivers in Central Norway affected by current or past mining operations. They found that only mayfly larvae (5-100 gm dry sample) were good potential biomonitors for Zn and showed promise for Cd and Cu in waters contaminated with moderate to low levels of the metals studied.

In addition to serving as monitors of the impact of potentially toxic metals in an ecosystem, faunal data can establish risk areas for deficiencies and excesses in feeding areas which comprise their ranges. On land, moose (livers, kidneys) and reindeer have been used as monitors. Selinus (1995) reported that a map of the Cu/Mo ratio in biogeochemical samples in an area of southern Sweden defined the risk areas for toxicity in ruminants from ingestion of excess amounts of Cu. The same map identified areas where toxicity could develop from Cu deficiency. The deficiency locations coincided with the highest prevalence of an unknown moose disease. It is possible that liming of lakes, fields and pastures and to a lesser extent forests was responsible for the Cu deficiency. Liming increases the pH of a growth environment. Higher pH values increase the mobility of Mo and increased Mo in feed inhibits the absorption of Cu by ruminants, thus causing a deficiency in this important nutrient. Webb (1971) reported simiar findings for cattle in England which suffered hypo-cuprosis because a high concentration of Mo in forage prevented the absorption of Cu from the forage.

Humans

Bioaccumulation and health impacts of metals such as Pb, As, Hg, Se and Cd on humans can be monitored by analysis of samples from humans. The monitoring establishes background and higher than background levels of potentially toxic metals. These data can also reveal ties to past and present incidences of diseases and hence project the health status for future populations.

For example, neurological problems in young children as a result of ingestion of Pb is well-documented. The Pb originates from dust in homes that have Pb-based paints or from soils that are mixed with paint scrapings. The established pathway is a hand-to-mouth transfer from soils that children play on (Thornton, 1993; Watt et al., 1993). Respiration of automobile or smelter emissions adds to the ingestion of Pb and other heavy metals by human populations. Comestibles that bioaccumulate metals from contaminated soils add to a pollution burden.

Monitoring the Pb content in urine and blood samples reveals those in a population who have bioaccumulated Pb (or other metal toxins) beyond maximum permissible concentrations set by health authorities. In the area of the Bukowno smelter in Poland, topsoils are contaminated with Pb, Zn and Cd (Verner and Ramsey, 1996). The level of Pb in blood samples from 75% of the children in the 5–8 year old age group was greater than the poisoning concentrations (>10 µg/dl) and the Cd contents in urine were 0.85 µm/l which far exceeded the health hazard limit of 0.5 µg/dl (Fergusson, 1990). Diseases attributed to intoxication from smelter sources include saturnine encephalopathy (slowly developing brain disease), radial nerve paralysis and saturnine colic. When sources of contaminants (e.g., smelters) are eliminated or their input alleviated and remediation of contaminated media (e.g., soils) proceeds, biomonitoring with blood and urine, for example, can reveal any improvements that are being made in what were hazardous health environments. Biomonitoring is going on where similar Pb and Cd intoxication was found in human populations living in the Baia Mare and Copsa Mica smelter areas in Romania. Of the 14,554 people sampled, almost 1200 had harmful levels of Pb (>250 µg/l) in their urine and more than 7500 had harmful levels (>150–250 µg/l) of Pb in their urine (Lacatusu et al., 1996). Here, as in Bukowno, young children were most susceptible to Pb and Cd poisoning.

Kido et al. (1991) studied Cd intoxication in 1843 exposed and 240 non-exposed inhabitants of the Kakehashi River Basin, a Cd-polluted area in Japan, in relation to the incidence of metallothioneinuria, a renal tubular dysfunction. Cadmium concentrations in rice were used as a reasonable measure of exposure. Kido et al. (1991) found a dose-related increase in metallothioneinuria. The critical total exposure was 2 g for men and women. Thus, a cumulative dose of 2 g over 50 years means an average daily intake of 110 µg, values that can be taken as maximum allowable lifetime and daily intake limits for chronic dietary exposure to Cd. This can be monitored in a population using urine analysis to complement rice grain analysis.

Following the research results from Kido et al. (1991), Nordberg et al. (1996) studied Cd in human urine as indicator of renal accumulation of Cd and its relation to renal disfunction. Rice and urine samples were collected from three areas in Zhejiang province, China, representing a highly-exposed Cd area, a medium-exposed area, and a control area. The Cd in rice for the three areas, respectively, was 3.70 ppm, 0.51 ppm and 0.072 ppm. This was paralleled to some degree by the Cd in urine which had geometric means of 10.7, 1.62, 0.40 µg/l, respectively, in the three areas. Nordberg et al. (1996) found a close response between renal dysfunction, Cd in urine, and ingestion of Cd in rice in the heavily-exposed group, thereby confirming earlier research.

There are different exposure levels for Cd (and other metals) over time which cause other diseases to develop. For example, In a review of Cd studies in Japan, Tsuchiya (1978) found that there was an average daily dietary intake of Cd from rice of about 350 µg in polluted zones where life-threatening osteomalacia or the itai-itai disease (cited in Chapter 2) was reported. On the basis of other studies on the relation between Cd ingestion and human disease, Kim and Thornton (1993) investigated the dietary intake of Cd on the Deog-Pyoung uraniferous black shale area in Korea. Black shales have high contents of several potentially toxic metals (Chapter 2) which can translocate to soils that form from them. Food crops may incorporate these heavy metals in their edible or useable portions. The concentration of Cd in rice grain was analysed at 0.6 µg/g for an average daily dietary intake from this rice up to 344 µg, an amount comparable to that from the itai-itai disease areas in Japan and more than 3 times that which was related to renal tube dysfunction cited above. The potential health problem is further exascerbated in this area from the consumption of the staples red pepper (1.9 µg Cd) and lettuce (2.3 µg) and from using tobacco (2.6 µg or 42 µg per pack) grown on the black shale derived soils. Biomonitoring of Cd in urine should be a standard health authority function here.

It is obvious from the cases cited above for Cd, that the amount of ingested Cd has to be diminished. This is true for other heavy metals as well (e.g., Pb) that are demonstrably damaging to ecosystem inhabitants. This may be possible through chemical remediation techniques of soils but can probably be better achieved using genetic modification technology so that rice grain (and other crops) discriminates against uptake of excessive concentrations of Cd and/or other potentially toxic metals that are present in agricultural areas.

In special environments such as the Arctic, seabirds, seals and polar bears show higher than normal hepatic concentrations of metals such as As, Cd and Hg. These metals originate on land or in the atmosphere and intrude Arctic seas. They access the marine food chain in plankton, algae and macrophytes and from suspended sediments and bioaccumulate in some fish and shellfish. The seabirds and seals feed on fish and shellfish and likely derive contaminants from them. Polar bears feed on ringed seals which have bioaccumulated metals. These translocate to the polar bears.

Many different sample types are analysed in environmental geochemistry research into the presence of potentially toxic metals in an ecosystem and its health status. There are preferred protocols for collecting and preparing different samples types and preferred analytical techniques for each element. These are discussed in the following chapter.

CHAPTER 7

Analysis of Indicator Media: Samples/Protocols

Geochemical analysis of sample types evaluated in environmental research (Table 6-1) should give accurate and precise results at concentrations close to detection limits as well as at high concentrations. Analytical techniques are chosen to obtain optimal results. This improves the basis for assessment and scientific interpretation of geochemical data. Nonetheless, more complete resolutions to environmental geochemistry problems require other data and considerations. A focussed research program starts with sample selection (Chapter 6) and a thorough description of measured and observed physical-chemical-biological parameters at sampling locations. This is followed by care in sampling, proper sample treatment and preservation in the field, and correct sample handling during transport to the laboratory and storage there. Next there must be care in preparing a representative sample for analysis in the laboratory, preparing replicates to monitor precision, and selecting standards matched to sample matrix type(s) and projected concentration ranges of target elements or compounds to monitor accuracy. The July 1994 (Vol. 18) Special Number of the Geostandards Newsletter lists producers of about 400 geochemical reference materials available to the analytical community. It describes the standards and provides data such as "best values" for the concentrations of about 60 chemical elements and also considers problems that might be associated with use of specific standards.

Analytical Choices

The analytical technique is chosen in order to minimize interferences that can affect an accurate and precise data output necessary to achieve the aims of an environmental research program. Preferred methodologies used in many current studies in the living environment include instrumental neutron activation analysis (INAA), inductively coupled plasma optical emission spectrometry (ICP-OES), inductively coupled plasma-mass spectrometry (ICP-MS), X-ray fluorescence spectroscopy (XRF), and flame, flameless and cold vapour atomic absorption spectrometry (AAS, GFAAS, CV-FIMS). The first three techniques have the advantage of being multielemental so that major rock-forming elements and environmentally important trace metals can be determined simultaneously. The ICP-MS method has an advantage over the ICP-OES method because it achieves lower detection limits, a feature important for analysis of natural waters. Selected analytical techniques preferred for determinations of specific potentially toxic metals in solid inorganic earth materials and their individual "working" detection limits are listed in Table 7-1. Table 7-2 gives the metals' "working" detection limits for vegetation analysed using INAA, ICP-OES, ICP-MS or CV-FIMS and for natural fresh waters analysed by ICP-MS.

The choice of an analytical technique based on manufacturers' specifications on detection limits can be a problem. These "best" detection limits are established from analyses on pure samples without an interfering matrix or interfering associated elements. Upper concentration limits that give a linear response to chemical signal will likewise be affected by these interferences. Matrices in environmental samples can be quite different. For examples, analyses may have to be made on silicates, carbonates, sulfides, organic matter, and other matrices, or on fresh and marine waters and brines. Each can cause a different response for the chemical signal which translates to an element concentration. Because of this the lower and upper concen-

Table 7-1. Some preferred techniques for the analysis of potentially toxic metals in inorganic solid geochemistry samples used in environmental studies (compiled from commercial laboratories literature, 2000). Detection limits in ppm unless otherwise noted

Element	Some Preferred Techniques		Detection Limits		Sample State
	Lab A	Lab B	Lab A	Lab B	
As	INAA		0.5		Solid
	AAH			0.1	Hydride
Be	ICP		2	0.5	Dissolution
Cd	ICP		0.5		Dissolution
		AA		0.2	
Co	INAA		1		Solid
		ICP		1	Dissolution
Cr	INAA		5		Solid
		XRF	5		Solid
Cu	ICP		1	1	Dissolution
Fe	INAA		0.01%		Solid
		ICP		50	Dissolution
Hg	CV-AA		5 ppb	5 ppb	Cold Vapour
Mn	ICP		1	2	Dissolution
Mo	ICP		2	1	Dissolution
Ni	ICP		1	1	Dissolution
Pb	ICP		5	2	Dissolution
Sb	INAA		0.1		Solid
		AAH		0.1	Hydride
Sc	INAA		0.1		Solid
		ICP		0.5	Dissolution
Se	INAA		3		Solid
		AAH		0.1	Hydride
Ti	ICP		0.01%		Dissolution
		XRF		5	Solid
V	INAA		2		Solid
		ICP		2	Dissolution
Zn	ICP		1	1	Dissolution

INAA – Instrumental Neutron Activation Analysis; AAH – Atomic Absorption Hydride Generation; ICP – Inductively Coupled Plasma, Atomic Emission Spectrometry (ICP-AES) = Optical Emission Spectrometry (ICP-OES); AA – Atomic Absorption; CV-AA – Cold Vapour Atomic Absorption; XRF - X-Ray Fluorescence.

Table 7-2. Detection limits of potentially toxic metals in ppm determined in vegetation samples analysed by a preferred technique by Lab A as noted. Lab B vegetation data are by INAA alone. Natural water samples values in ppb as analysed by ICP-MS. (Compiled from commercial laboratories literature (2000)

Element		Detection Limit			
	Preferred Technique Lab A	Vegetation		Natural "Fresh" Water	
		Lab A	Lab B	Lab A	Lab B
As	INAA	1 – 0.01	0.01	0.03	0.1
Be	ICP-MS	0.01	ND	0.1	0.1
Cd	ICP-OES	0.1	0.5	0.01	0.01
Co	INAA	0.1	0.3	0.005	0.1
Cr	INAA	0.3	0.3	0.5	0.1
Cu	ICP-OES	0.5	ND	0.2	0.1
Fe	INAA	50	0.005%	5	ND
Hg	CV-FIMS	0.005	50 ppb	0.006	0.2
Mn	INAA	0.01	ND	0.1	0.1
Mo	INAA	0.05	0.05	0.1	1
Ni	ICP-MS	1	5	0.3	0.1
Pb	ICP-MS	1	ND	0.1	0.01
Sb	INAA	0.005	ND	0.01	0.1
Sc	INAA	0.01	0.2	1	0.1
Se	INAA	0.1	0.5	0.2	0.1
Ti	ICP-OES	2	ND	0.1	ND
V	ICP-MS	0.1	ND	0.05	0.1
Zn	ICP-MS	1	2	0.5	0.1

ND – not determined.

* Detection limits for ocean water or natural brines will be greater by an order of magnitude.

INAA   – Instrumental Neutron Activation Analysis.
ICP-MS – Inductively Coupled Plasma-Mass Spectrometry.
ICP-OES – Inductively Coupled Plasma-Optical Emission;
             Spectrometry = Atomic Emission Spectrometry.
CV-FIMS – Cold Vapour-Flow Injection Hg System.

tration limits that can be accurately and precisely determined with an "ideal" methodology may not be those achievable and necessary for environmental investigations. The concept of "best" detection limit is realistically replaced by that of the functional or "working" detection limit.

Systematic errors are from unknown or unexpected factors such as in field sampling, impurities in reagents, a poor match of standards/matrix with samples, or partial solubility of precipitates. Random errors are those that develop from factors such as instability of electronic components, sample weights that are slightly wrong, and volumes that are not exact. Both systematic and random errors diminish the reliablity of analytical data and are of great concern to environmental geochemists. These can be monitored by random numbering of samples in the field, reordering samples prior to laboratory preparation, and then randomizing again previous to the analysis. Geochemical reference standards should be prepared and analysed along with samples from a study suite for an additional control on the laboratory/analytical phase of a study and, of course, on accuracy and precision of results.

Bulk/Total Sample

In the previous chapter, the many classes of samples used in environmental research were evaluated and the benefits and limitations of using each assessed. In many programs a bulk sample is analysed for its potentially toxic metals content. The resulting data may demonstrate that some metal values are elevated but to levels that fall within natural variations in baseline or background values. As discussed in Chapter V, the limits (range) for a normal distibution that brackets the natural concentration fluctuations in a population representing a geographic region could be set as the arithmetic mean concentration of a metal (X_a) ± one standard deviation (σ). For geochemical distributions which often are lognormal, the range would likely be more accurate if defined by the geometric mean value (X_g) ± one standard deviation (b). Metal baseline values for bulk samples calculated as above can be misleading for environmental geochemistry evaluations, although mathematically correct. This is because of the diluent effect on metal contents in a bulk sample by components that bear little of metals being assessed. Bulk sample analysis baselines can conceal a high toxic metal concentration in a sample component from which the metal is bioavailable to a food web. The questions then are what mineral or size fraction in bulk samples concentrates potentially toxic metals and how bioavailable are they to an ecosystem.

This is a common case in soil samples in which potentially toxic metals concentrate in a B-soil horizon clay-size fraction. Food crop roots tapping the soil for nutrient elements may absorb metals in proportion to their concentrations in the soil horizon, or may accumulate or hyperaccumulate the metals, and translocate them to an edible part (Table 6-4). Over time this can lead to a bioaccumulation of metals to toxic levels in the food crop consumer. Conversely, if a food crop discriminates against the uptake of a essential micronutrient metal essential to a consumer, this can lead to a deficiency in a diet if the metal is not otherwise available in the food chain. Finally, heavy metals may be in components of a soil horizon from which they are not bioavailable. Thus, bulk sample geochemistry may not reflect the real risk to an ecosystem from potentially toxic metals. Selective subsampling and/or selective extraction is necessary to clarify bioavailability from a source.

Selective Sampling

By Size Fraction of Inorganic Media

Selective subsampling of inorganic solids such as soil or sediment can involve a particular stratum in a soil or sediment core and/or a fine-size fraction comprised of silt ($< 63 \, \mu m - > 3.9 \, \mu m$) and clay ($< 3.9 \, \mu m$ – colloid) in the soil horizon or sediment layer. Metals concentrate in the clay-size sediment fraction more than other size fractions by adsorption onto charged surfaces of minerals and associated amorphous solids. They can also be absorbed into solid matter. Those phases with the greater surface area (e.g., in m^3/gm) will have the highest sorption capacity (Table 7-3). The metals are bonded to the sediment with varying strengths thereby affecting their bioavailability in ecosystems. Under one set of physical, chemical and biological properties a metal may be readily mobilized yet be immobile under a different set of ambient conditions. Table 7-4 illustrates the

Table 7-3. Typical surface areas of sorbents that tend to decrease metal solubility (Bourg, 1995)

Soil minerals and soils	Surface area (m^2/gm)
Smectite (aka montmorillonite)	700 – 800
Illite	65 – 100
Kaolinite	7 – 30
Mn oxides	30 – 300
Fe oxides	40 – 80
Clays and loams	20 – 270
Silty loams	20 – 200
Rendzinas	15 – 170
Sandy loams and loamy sands	10 – 70
Calcareous aquifer sand	0.5 – 5

Table 7-4. Potentially toxic metals' mobilities in supergene (e.g., soil) environments with different pH and oxidizing-reducing conditions (from Siegel, 1992)

Relative Mobility	Oxidizing, pH 5–8	Oxidizing, pH < 4	Reducing
Very mobile	Mo (Se)		
Moderately mobile	Zn, V, As (Hg, Sb)	Zn, Cd, Hg, Cu, Co, Ni, V, As, Mn	Mn
Slightly mobile	Mn, Pb, Cu, Ni, Co, (Cd)		Fe
Immobile	Fe, Sc, Ti, Sn (Cr)	Fe, Sc, Ti, Sn, As, Mo, Se	Fe, Ti, Sn, Cu, Pb, Zn, Cd, Hg, Ni, Co, As, Sb, V, Se, Mo, Cr

relative mobility of potentially toxic metals with varying conditions of pH and Eh in a supergene (e.g., soil) environment. This parameter of mobility or immobility of a metal, whether or not a micronutrient, can result in an environmental intrusion that is economically damaging. For example, Mo will be immobile in acidic soils but if liming is used to condition soils to basic pH levels, Mo becomes bioavailable and may be detrimental to ruminants. An excess of Mo in soils taken up in forage will interfer with the absorption of the nutrient Cu in cattle. This leads to a Cu deficiency and the disease hypocuprosis. In dairy cattle hypocuprosis causes weight loss, a degradation in animal condition, a drop in procreation and a decrease in milk production (Webb, 1971). Differences in soil pH that stimulate metal mobility as just described and create a disease condition can be natural as well as anthropogenic. As a matter of protocol, pre-development research into the physical, chemical and biological conditions in an ecosystem and the mobility/immobility of metals can be done to insure that the productivity (economic viability) of a planned project or group of projects can be maintained over the long term.

Suspended Matter

Suspended sediments (generally taken as $> 0.45\ \mu m$ in size) are transported by fluvial, lacustrine, estuarine or marine waters and by atmospheric currents. Suspensates may carry high concentrations of potentially toxic metals from natural and/or anthropogenic sources described in Chapter II. During transport from source to depositional environment, metals in suspended matter can enter an ecosystem food web if bioavailable to life forms that ingest or use the fine-size matter as a nutrient source.

Suspended particulates are separated from water for chemical and mineralogical analysis plus mass per volume determinations using special filters (e.g., Nucleopore® or Millipore®) and vacuum-filter equipment. Because the suspensate mass is generally low, it may be necessary to filter large volumes of water to obtain the sample weight necessary for an analysis. In the marine environment or deep lakes, samples can be collected at different depths such as surface, intermediate and deep waters. The suspensates are collected using Niskin bottles with a 30 liter capacity. These bottles can be clustered to collect greater water volumes in order to obtain the sample weight required for chemical analysis. Analysis by ICP-OES or ICP-MS, for example, ideally requires 0.5 g of sample but smaller masses can be used as well. The sample is recovered from a filter by washing it off or by filter dissolution or wet ashing. In the latter two cases, corrections are made to account for metals that are in the filters. Suspended sediments in streams and rivers, in groundwater, and in the atmosphere are collected *in-situ* using powered portable collection and filter systems or a manual unit (Siegel, 1985). Fixed filter capture systems or airborne collectors are used to obtain samples of suspended matter from the atmosphere near a land surface or from high altitudes. Deposition of suspended matter can convey high metal contents to terrestrial, estuarine, and oceanic sedimentary environments and food webs. As emphasized in previous chapters, bioaccumulation and biomagnification along a food chain and in a food web can threaten organisms' health.

Water

Water filtered free of its entrained solid load in the field or soon thereafter (commonly with 0.45 µm filters) should be acidified with 1 ml/500ml of concentrated ultrapure nitric acid. The acidification can not be done before suspended solids are removed because it may cause a release of metals to a water sample. High contents of potentially toxic metals in water from aquifers or springs and from streams can be traced upflow to natural sources or to metal-bearing effluent discharges. Similarly, contamination of lake/pond waters by natural and/or anthropogenic sources can be traced to inflow zones whether from streams or from the subsurface. The input location of pollutants identified in sea water samples may be traceable using a knowledge of marine currents.

Very low detection limits in natural water samples with < 500 ppm of total dissolved solids can be achieved using ICP-MS for the analysis (Table 7-2). Water samples with > 500 ppm total dissolved solids, marine waters, and brines may have greatly elevated detection limits unless the matrix effect is attenuated or eliminated. Estuarine waters transition from fluvial and groundwater discharge to sea water with 35000 ppm total dissolved solids. Mixing of the two waters can be slow especially close to the surface depending on the discharge of rivers, depth of the estuary and tidal reaches. If dilution of pollutants is slow, metals found by analysis of land discharge can be detected at a good distance from shore. In this case, the zone of influence of toxic metals that can enter the estuarine food web and affect organism growth and productivity there can be delimited. This is important to environmental planning with respect to any development in river catchments because estuaries are the spawning grounds for a large percentage of commercial marine foodfish. Toxic metals in the ecosystem can affect spawning and hence ocean fisheries. In addition, fish and shellfish in estuarine fisheries are a regionally important food supply. However, a harvest can suffer where stocks are at risk from extractable toxic metals in the ecosystem.

In contrast, the sea water reservoir is a diluent and does not lend itself readily to the study of heavy metals pollution in the medium except for special conditions. Water samples from outfalls where untreated or incompletely treated industrial effluents discharge into coastal waters pollute the ecosystem. Foodfish can bioaccumulate metals from input at an outfall and pass them up the food web. Sustained disposal of toxic metals-bearing wastes offshore may likewise lead to the release of the metals and their access to the foodweb. Consumption over time of foodfish contaminated by toxic metal(s) can lead to disease and death in human populations such as happened at Minamata Bay, Japan (see following Chemical Species section for discussion).

Water samples from deep-sea vents (e.g., "black smokers") where hydrothermal waters meet sea water have high contents of heavy metals. These form metal-rich sediments or precipitate as massive sulfide deposits. Because of the ambient stability of the precipitated minerals, limited mobilities of metals from them, and the depths of the waters into which the vented metals-bearing fluids are released, the metals pose little or no threat to ocean fisheries or deep-sea ecosystems.

Analysis of Organisms

Analysis of organism parts can indicate the etiology of a disease or abnormal conditions in a population when clinical symptoms are diagnosed and toxic metals may be involved. Table 7-5 gives some examples of organs adversely affected by some of the potentially toxic metals and some resulting health effects. Analytical protocols have been established for different samples from organisms that are used to monitor and define the health of an environment. International and national organizations such as WHO, the EPA, and the EC have published these protocols which must be followed precisely to be accepted for decision making.

The toxic metals' involvement is generally long-term and can cause disease in three ways. One is through ingestion of an excess concentration of an essential element or a non-essential element. A second is from a deficiency in the intake of an essential micronutrient. The third is through a condition brought about by intake of a metal which inhibits the absorption of an essential element by an organism. Analyses are made of vital organs (e.g., livers, kidneys, muscles), growing parts (e.g., twigs, leaves, hair, nails, antlers), and body fluids (e.g., blood, urine). A causal relation between a disease or abnormal condition in a living population and metal contents alerts health scientists to the possibility that similar populations may be at risk from essential element excess or deficiency when only subclinical symptoms exist. If the former condition is defined, the source(s) of excess concentrations of the essential yet potentially toxic metals in an ecosystem can, in many cases, be determined and its release into the living environment mediated. If deficiency concentrations are found, it is possible to amend or condition an ecosystem so that essential elements become available and extractable from a growth environment. Finally, a therapy can be developed which ameliorates and eliminates a disease threat to a population.

Table 7-5. Examples of organs and systems impacted by the bioaccumulation over time of toxic concentrations of metals/metalloids (after Zielhuis, 1979)

System/Organ	Toxic Element(s)	General Health Effects
Central Nervous System	CH_3Hg^+, Hg Pb^{2+}	Brain damage Reduced neuropsychological functions
Renal System	Cd	Tubular, glomerular damage, proteinuria
	Hg^{2+}	Tubular nephrosis
	As	Tubular disfunction
Cardiovascular System	Cd As	
Reproductive System	CH_3Hg^+, Hg As	Spontaneous abortion
Blood System	Pb	Inhibits biosynthesis of haem
	Cd	Slight anemia
	As	Anemia
Respiratory Tract	Cd	Emphysema
	As	Emphysema and fibrosis
	Hg	Broncial effects
	Se	Respiratory inflammation
Brain and body	CH_3Hg^+, Hg	Deformation
Liver	As	Cirrhosis
Prostate gland, lung	Cd	Cancer
Skin, lung	As	Cancer
Skeleton	Cd	Osteomalacia ("itai-itai")
	Se	Tooth caries
Chromosomes	Cd, As	Aberration

Selective Extraction

Selective chemical extraction of potentially toxic metals from soils or sediments provides insight on how the metals are incorporated in various component phases, and from this, their bioavailability. Table 7-6 presents several selective extraction procedures that have been described in literature reports since the 1970s. Selective extraction generally follows a sequential procedure that is designed for each matrix. This gives a high discrimination threshold towards a fraction present and also avoids significant interferences among inorganic and organic components. The procedures for analysing heavy metals in different matrices of earth materials are constantly being reviewed, modified, and newly developed to improve precision and accuracy for a target phase and elements bound to it. During 1979, Tessier et al. presented a flowchart for a sequential extraction scheme (Figure 7-1 A). Breward and Peachey (1983) modified the sequence to distinguish between metals tied up by different organic fractions and also to separate elements associated with secondary manganese oxides from those in secondary iron oxides (Figure 7-1 B). Hall et al. (1996) modified the sequential schemes so as to distinguish between elements bound to amorphous iron oxides and those found in crystalline iron oxides as well as accounting for elements bound to a sulfide phase (Figure 7-2). The text accompanying the figure from Hall et al. (1996) details a 22 step procedure to be followed in the extraction sequence. Gray et al. (1999) followed a similar selective and acid extraction procedure in which elements associated with Mn oxides were determined but those in sulfides and organics were not (Figure 7-3).

Results of element distributions in sample fractions are reported in terms of multiple components. In sediments these components may be elements adsorbed or in exchange positions, in carbonates, in authigenic minerals, in hydrogenous or lithogenous phases (Fe and Mn amorphous or crystalline oxy/hydroxides), in phases that are difficultly, moderately or easily reducible, in organics, in sulfides, and

Table 7-6. Summary of extraction schemes in the literature (after Hall et al., 1996)

Exchangeable	Adsorbed, carbonates	Mn oxides	"Soluble" organic phase	Amorphous Fe oxyhydroxides	Crystalline Fe oxides	Sulfides and organics	Residual (silcates)	Ref.
	NaOAc, pH 5				DCB	H_2O_2	Separated by size fraction	1
		0.1 M $NH_2OH \cdot HCl$ in 0.01 M HNO_3		0.25 M $NH_2OH \cdot HCl$/ 0.25 M HCl	DCB	$KClO_3$-HCl HNO_3	HF-HNO_3	2
1 M NH_4OAc at pH 4.5		0.1 M $NH_2OH \cdot HCl$/ 1 M NH_4OAc		$NH_2NH_2 \cdot HCl$/ HCl, pH 4.5		H_2O_2/ NH_3OAc	HF	3
1 M $MgCl_2$ at pH 7	1 M NaOAc, pH 5		NaOCl, pH 9	0.04 M NH_2OH · HCl in 25% HOAc		H_2O_2-HNO_3	HF-$HClO_4$	4
		0.1 M $NH_2OH \cdot HCl$, pH 2.5		0.175 M $(NH_4) C_2O_4$ in 0.1 M $H_2C_2O_4$	DCB		HCl_4-HNO_3	5
1 M HOAc		0.1 M $NH_2OH \cdot HCl$ in 0.01 M HNO_3		0.25 M $NH_2OH \cdot HCl$ in 25% HOAc		H_2O_2/ NH_4OAc in 6% HNO_3	HF-HNO_3-HCl	6
1 M NH_4OAc at pH 4.5		0.1 M $NH_2OH \cdot HCl$, pH 4.5		0.175 M $(NH_4)C_2O_4$ in 0.1 M $H_2C_2O_4$, pH 3.3	0.175 M $(NH_4)C_2O_4$ in 0.1 M $H_2C_2O_4$ under UV	35% H_2O_2	HF-HCl	7
		0.1 M $NH_2OH \cdot HCl$	NaOCl	0.25 M $NH_2OH \cdot HCl$ in 0.25 M HCl	1.0 M $NH_2OH \cdot HCl$ in 25% HOAc		HNO_3	8
1 M NH_4OAc at pH 4.5		0.1 M $NH_2OH \cdot HCl$ pH 2.5		0.175 M $(NH_4) C_2O_4$ in 0.1 M $H_2C_2O_4$, pH 3.3, in dark	0.175 M $(NH_4) C_2O_4$ in 0.1 M $H_2C_2O_4$, under UV		HF-HNO_3-HCl	9

DCB: Dithionite citrate buffer. 1: Rose and Suhr (1971); 2: Chao and Theobald (1976); 3: Gatehouse et al., (1977); Tessier et al., (1979); 5: Hoffman and Fletcher (1979); 6: Filipek and Theobald (1981); 7: Sondag; 8: Bogle and Nichol (1981); 9: Cardoso Fonseca and Martin (1986).

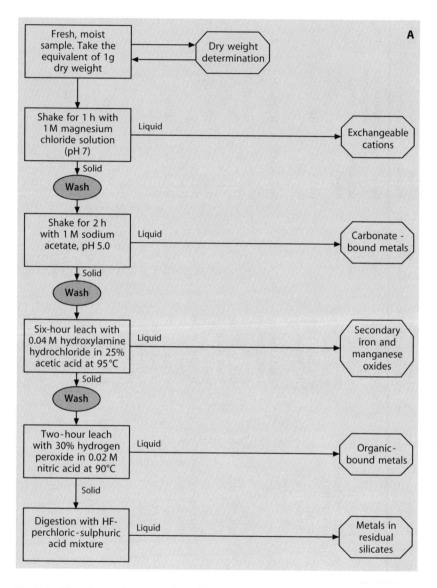

Fig. 7-1. Flowcharts for sequential extraction schemes. **A** After Tessier et al. (1979)

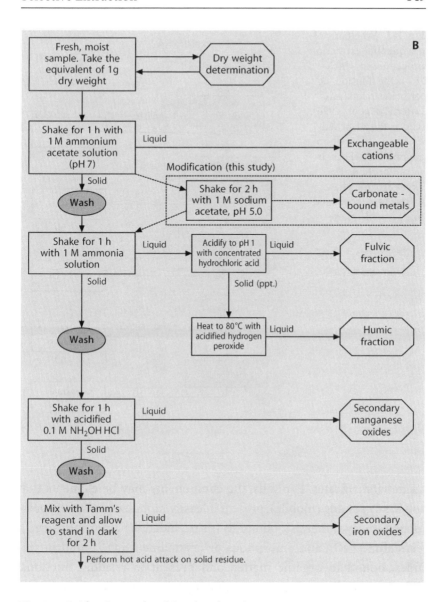

Fig. 7-1. B After Breward and Peachey (1983)

Fig. 7-2. Schematic of a sequential extraction procedure in which "F" is the dilution factor prior to analysis (after Hall et al., 1996). AEC is the adsorbed-exchangeable-carbonate phase; AmFeOx is the amorphous Fe oxyhydroxide phase; CryFeOx is the crystalline Fe oxide phase

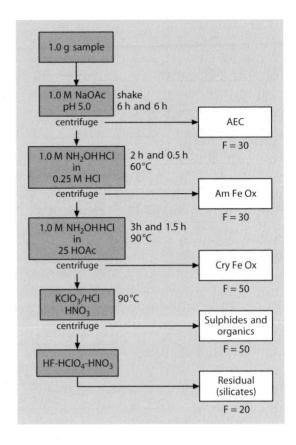

in detrital silicates. For soils, the components may be elements that are easily soluble (mobile), present in easily mobilized exchange positions [of clay minerals], bound to (or occluded in) amorphous FeO_x, crystalline FeO_x, and amorphous or crystalline MnO_x, found in sulfides, bound to organic matter, and present in residual fractions. Selective solubilization modes that can be used for each of the phases given above are listed in Table 7-6 or in the sequences given in Figures 7-1, 7-2 and 7-3. The most bioavailable metals are in the easily soluble phases and in ion exchange or adsorbed sites. The least bioavailable elements are in the residual or detrital silicate phases.

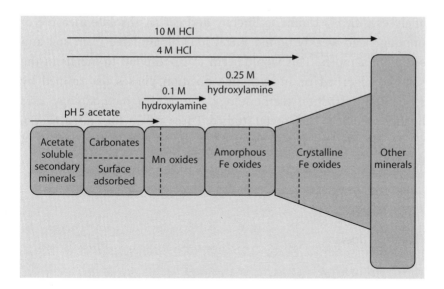

Fig. 7-3. Diagrammatic representation of the phases dissolved by selective and acid extraction (after Gray et al., 1999)

As noted in the previous chapter and Figure 6-3, the element partitioning (of Cd, Cr, Cu, Mo, Ni, Pb and Zn) in soil phases from Poland determined by selective extraction illustrated that Cd is the most bioavailable metal whereas Cr is least bioavailable. The relations between extractability of an element from solid phases in natural environments will vary with physical (e.g., texture), chemical (e.g., pH, redox potential) and biological (e.g., microbial activity) conditions in particular settings. Descriptions of selective/sequential extraction methods can be found in publications by (Tessier et al., 1979; Chester and Hughes, 1967; Breward and Peachey, 1983; Chao, 1984; Bendell-Young et al., 1992; Hall, 1992; Breward et al., 1996; Hall et al., 1996; Hall and Pelchat, 1997; Hall and Bonham-Carter, 1998).

Selective extractions focus on techniques that attack specific substrates in soils and sediments. This does not give a total value for metals that may be mobilized from all the substrates without following a complete selective sequential extraction protocol. Mobile metal ion analyses (MMI®) claims to liberate and get an absolute measure of unbound and weakly bound metal ions bioavailable from soil and sediment samples. This would seem to correspond to metals in the soluble, adsorbed or in exchange position phases determined by sequential extraction methodology. The MMI® methodology uses ligands specific to each metal studied. These are Cu, Pb, Zn, Cd, Ni, Co, Cr, As, Hg, Sb, Mo, Se and Fe or 13 of the 18 metals reported on in this text.

Chemical Species

Reuther (1996) defines a chemical species as being a molecular arrangment of a particular metal or metal compound. Chemical species are natural and man-made. They may be thermodynamically stable as free aqueous ions, as distinct dissolved inorganic salts or acids, or as solid oxides, sulfides and carbonates. Thermodynamically unstable aggregates such as dissolved and solid organic complexes, colloids, or co-precipitates, form another class of chemical species.

Speciation of a heavy metal in its physical, chemical, biological environment affects its extractability and mobility. Chemical speciation affects a metal's capacity to bioaccumulate in living things and hence its toxic effect. The speciation may be biomediated (Table 7-7) or may be the result of changes in the physical-chemical environment. For example, methyl mercury (CH_3Hg^+) is the most toxic species of Hg. Mercury may enter an ecosystem as the free ion, charged ion, or as a complex such as $HgCl_3^-$. In the marine environment, for example, Hg species can be absorbed by plankton, transformed to CH_3Hg^+, and then ascend a food web as larger marine organisms ingest the smaller ones. Along the food web, organisms may bioaccumulate the toxic Hg species to levels greater than the World Health Organization

Table 7-7. Micro-organism mediated toxic metals oxidation-reduction and bio-methylation-demethylation reactions (modified from Doelman, 1995)

Reaction	Toxic Metals Affected
Reduction	As^{5+}, Cr^{6+}, Fe^{3+}, Hg^+, Mn^{4+}, Se^{4+}
Oxidation	As^{3+}, Cr^{3+}, Fe^0, Fe^{3+}, Mn^{2+}, Sb^{3+}
Biomethylation	As^{5+}, Cd^{2+}, Cr^{2+} Hg^{2+}, Pb^{2+}, Se^{4+}, Sn^{2+}
Demethylation	RHg^{2+}, R_3Pb^+

maximum permissible level (MPL). For Hg in foodfish, this is 0.5 parts per million (0.00005 % by weight). At the top of the food web, human consumption of foodfish with greater than maximum permissible levels of Hg over time can lead to a bioaccumulation in consumers and the development of clinical symptoms of Hg poisoning. Ultimately, Hg poisoning can lead to disability, incapacitation and death as was the case at Minamata Bay in Japan during the 1950s. Here inhabitants of a coastal village had fish as a principal food source. An industrial plant discharged inorganic Hg into the bay waters. The Hg^{2+} (98%) was converted by biomethylation to methyl mercury which was accumulated by organisms at low trophic levels. This was biomagnified by fish along the food web to contents of 50 ppm (0.005%). Over about five years, Hg from ingestion of fish bioaccumulated in the human population to the degree that subclinical symptoms of Hg poisoning gave way to clinical symptoms. The effects of Hg poisoning on the nervous system can not be reversed. Forty-three persons died, 116 were permanently incapacitated. Others are living impaired by the poisoning. Control groups are being followed to determine if the Hg poisoning caused any genetic damage.

Demethylization by micro-organisms in soil, water and marine sediments, for example, can detoxify methyl Hg following the reaction $CH_3Hg^+ \rightleftarrows Hg^{2+} + Hg^0$. Conversely, demethylization or dealklation of tetraethyllead (CH_4Pb) will alter Pb to triethyllead (R_3Pb^+), the most stable and most toxic derivative of the metal.

International, Regional, National Protocols

Sampling, storage, preparation and analytical protocols have been established by environmental, health and safety departments worldwide for the analysis of potentially toxic metals that may be found in environmental samples. In the United States, for example, the U.S. Environmental Protection Agency (1992) described in detail 13 analytical methods covering 35 analytes that may be present in a variety of environmental sample types. De Zuane (1997) summarizes preferred EPA protocols on sampling, sample preparation and analytical techniques followed to determine water quality. The European Community, World Health Organization, and national health ministries plus provincial and state public health units have published like protocols that are followed to obtain optimal results on the determination of heavy metals in environmental samples. These include a range of waters from potable to surface, estuarine, marine and brines, industrial and municipal wastewater effluents, landfill leachates and groundwater. In addition there are methods that accomodate soils, suspended sediments, bottom sediments, dredged material, sludges, solid waste and biological tissues (e.g., edible fillet of fish).

Protocols differ somewhat among the environmental, health and safety departments but for the most part the analytical results on international standards are very much the same within the detection limits, accuracy and precision. They do change as technological advances are made that lower detection limits, extend the linear response of concentration to analytical signal, and improve on accuracy and precision. They change as well when new standards become available for analytical quality control.

CHAPTER 8

Remediation/Alleviation
of Metal(s) Contaminated Media

A major role of geochemistry in environmental projects is to assess clean-up possibilities for ecosystems that contain pollutant concentrations of potentially toxic metals that could access a foodweb. This role extends to remediation of environments where, in addition to heavy metals, extreme conditions occur that threaten ecosystem life such as low pH (acidic) waters or waters with limited BOD capacity. The targeted cleanup media include solids, liquids and gases from contaminated soils, groundwater and surface waters, sediment (fluvial, lacustrine, estuarine, marine), waste disposal sites and sewage sludge (industrial, agricultural, mining and municipal), and chimney emissions (e.g., smelting and electricity generating facilities).

A geochemist is educated in geological and chemical principles and processes. As such he/she has an added responsibility to highlight chemical characteristics of environments that may be useful to other researchers. This would include those working on prediction methods applied to natural physical geological hazards. Some hazards for which chemistry is being monitored for use in prediction are earthquakes (Rn and He emissions), volcanoes (SO_2 and CO_2 emissions), and landslides (Ca/Na in clay-rich soils). Continuing investigation of chemical conditions in the physical environment where such hazards now present a threat may lead to more reliable predictions of their onslaught. Also, a knowledge of changing chemical conditions with time and hazard events may help to develop methods that allow pulsed, small stress energy releases over time in a high risk hazard zone thus obviating a short-term full energy release.

Remediation technologies of heavy metals-contaminated environments are complex and may involve a combination of physical, chemical and biological methodologies. These have the purpose of immobilizing, containing, isolating, extracting and/or disposing of potentially toxic metals. Whatever is determined to be the best approach to remediation is predicated on cleaning up a polluted environment so that ecosystem inhabitants can safely use it. If this is not realistic within economic constraints, a short-term aim is to alleviate input from pollutant sources, and ultimately to further attenuate or eliminate them when economic resources, added knowledge and improved technology become available.

Soils

There are several factors that must be determined in an area before selecting a remediation methodology that is considered to be best suited for cleanup of heavy metals-contaminated soils and then initiating a pilot study. These include the extent and thickness of the polluted soils, soil texture (sand/silt/clay contents), porosity and permeability, organic matter content, mineralogy, chemical composition, and how target metals are bound in the soil (extractability – see Figure 6-3). Site specific data (e. g., mobility, kinetics of exchange between solid phase and solution) are needed to determine the best cleanup possibilities. These data will set the economic constraints of a remediation treatment/replacement/disposal total operation vs. long-term benefits that would be derived from a clean environment.

There are several options which can be considered for a cleanup to natural or close to natural conditions after the parameter assessments cited above are known. Some are physical and involve excavation and

replacement or capping and covering of metals-polluted soils. Some
are chemical and lead to pollutant extraction (flushing/washing) from
a soil or immobilization in a soil. Others are biological and are based
on vegetation (bacteria) that mobilize heavy metals for extraction
and vegetation that bioaccumulates potentially toxic metals from a contaminated soil or water. Each methodology has benefits and limitations with respect to the time for and completeness of contaminant
metal removal, cost, and any secondary (disposal) problems that have
to be part of mandated remediation requirements. Two critical factors
are the size of an area and the volume of matter to be remediated.

Excavation of thick masses of soils over an extended area is not
feasible economically. If the thickness of contaminated soil is not great
the cost may be borne nationally if there is extreme risk to a population and its ecosystem. After the Chernobyl nuclear disaster, the upper
few inches of soil contained dangerous levels of radioactivity over
many kilometers of what had been productive farmland north of the
blast site. The decision was made to skim off the top few inches and
store the radioactive material at a secure facility until the level of
activity dropped to acceptable levels or until permanent safe disposal
was available.

On the other hand, even if an area of metal-polluted soil is relatively small, < $^1/_2$ hectare (~ $1^1/_4$ acres) and the polluted section not deep
(< 1 m), fluid or electrolytic extractive techniques will not work effectively if the soil has low permeability (high clay content). Excavation
can be the remediation modus operandi. Replacement with clean
soil and reclamation of the topographic surface can make an area
reuseable for projects sensitive to soil chemistry such as truck farming
or flower growing. Realistically, however, the small area and high costs
of excavation and soil replacement plus treatment and disposal of
the excavated soil make it likely that the terrain would be used for
residential or industrial development for which cleanup investment
recovery is relatively fast.

Where texture of a soil imparts good porosity and high permeability, as in very sandy soils, acid or chelant extraction of pollutant

metals is possible by washing/flushing followed by capture, immobilization, containment and disposal. This presumes that the metal or metals to be mobilized are in/on extractable phases (Figure 6-3) when a pressure-driven infiltrating fluid moves from an injection well through a metal-contaminated phase and then to discharge wells. Extraction efficiency is enhanced *in situ* when conditions change to those that favor metal mobilization in an aqueous solution by increasing the solubility of a target metal or metals (Rulkens et al, 1995). This is mainly done by the addition of reagents such as acids or peptizers, or bacteria that change redox potential.

Urlings (1990) reported that Cd was extracted from soils in Holland using an acidic (HCl) solution at pH = 3.5. Cadmium concentration in soils decreased from 10 to 1 ppm at a cost that was less than if excavation had been used. The maximum permissible concentration for Cd in agricultural soils is 3 ppm. The extracted pollutant-bearing solution has to be treated to remove metals before its disposal. Other acid extractants that are efficient when they move through heavy metals-contaminated soils are acetic acid, citric acid, and di-Na EDTA. At the Aberdeen Proving Ground soil test site, for example, Peters (1999) determined that 70% of the contaminant Cr, Cu, Pb and Zn in exchangeable, carbonate and reducible oxide phases were amenable to soil washing/flushing whereas Cd, Fe and Mn were not as amenable. Chelant extraction was done using EDTA and citric acid. The chelant solution removed As, Cd, Cr, Cu, Fe, Hg and Pb simultaneously with a > 97% extraction of Cu, Pb and Zn.

Metal solubility in soils is rarely determined by the dissolution or precipitation of a single metal or compound but is influenced instead by associated metals or compounds involved in competing reactions. The conditions at soil remediation sites determine metals species that are stable there. Metal species react differently to washing/flushing chemicals so that leaching conditions are set to maximize extraction efficiency. For example. Thoming et al. (2000) studied the leaching of Pb and Hg in soils and determined that the optimum pH of the extractant solution should be between 5 and 6 and at specific redox conditions since soluble anionic complexing is different. Leaching of Pb with acetate and citrate solutions was 97.5% and 98% efficient, respectively. Leaching of Hg from the soil with electrolytically activated NaCl solution followed by rinsing with water had a removal rate up to 99.6%.

Although EDTA is a common chelator for the extraction of heavy metals from soil, it may not be effective in some soils. For example if the soil has a pH < 6, amorphous Fe in the soil will compete for EDTA ligand sites and the efficiency of extraction for several potentially toxic metals is diminished. In calcareous soils too much EDTA would have to be used to extract heavy metals present increasing the costs of remediation significantly. In this latter case a different extractant solution or combination of methodologies would be selected. In the end, specific reagents have been shown to be effective for extraction of metals from individual soil components.

Electro-remediation of metal-contaminated soil in a chemical extraction system improves the rate of migration of ions between a cathode and an anode. However, in some soil environments there is a precipitation and sorption of metals near the cathode that reduces the mobility of heavy metals and limits the cleanup process. In laboratory experiments, Wong et al. (1997) found that for Pb and Zn, the addition of EDTA to the catholyte solubilizes the precipitated or sorbed metals. The resulting complexes were moved to the anode with a removal efficiency close to 100% for spiked samples. Increased concentration and residence time of a chelant in soil will improve solubilization efficiency.

Low permeability (clay-rich) soils present a formidable barrier to the use of a chemical reagent wash/flush technique. An electro-remediation approach has promise (Monsanto, 1995). Initially, this involves the creation of highly permeable zones in a contaminated soil (e.g., by hydraulic fracturing). These are used as degradation/desorption pathways for injected agents such as microbes and chemicals. Electro-osmosis is then used as a liquid pump for flushing pollutants from the soil and moving them to collection and treatment facilities.

This is followed by switching electrical polarity thus reversing the fluid flow which both increases the completeness of pollutant removal and allows multiple passes of chemicals through treatment zones for increased degradation/desorption. The extracted pollutants have to be properly recycled or disposed of to prevent them from invading proximate ecosystems.

The chemical extraction process may be facilitated at specific sites by using microbial leaching as a method to solubilize metal contaminants in a soil and make them available to an acidic aqueous extractant phase. This presupposes that the essential micronutrients such as Co, Cr, Cu, Fe, Mn and Zn are available in the soil at concentrations that will not harm bacterial growth (Jackman and Powell, 1991). A potential microbial-assisted cleanup site has to be carefully evaluated because high concentrations of the micronutrients Cu and Zn or of As, Pb and CN⁻ can inhibit microbial enzyme formation and cause bacteria to die (Hasan, 1996). When microbial leaching is used to aid aqueous extractant remediation, factors that favor high mobilization are pH (best at ~7.0), temperature (between 15°C and 50°C) and sufficient moisture.

Another chemical methodology changes conditions *in-situ* so as to immobilize pollutant metals. Raising the pH of a soil by liming it can immobilize several metals. However, it may mobilize others. For example, under slightly acidic to neutral soil conditions, Mo is immobile. If soil pH is altered towards basic by liming, Mo will be mobilized. This becomes important in agricultural zones where livestock may not be able to absorb the nutrient Cu from forage in which it is plentiful because excess Mo in the forage prevents absorption of Cu. In the past, this caused hypocuprosis, a disease that afflicted the physical condition and productivity of dairy cattle (Webb, 1971).

Beringite is a modified aluminosilicate that originates from the fluidized bed burning of coal refuse from the former coal mine of Beringen in NE Belgium. Beringite has a high metal immobilizing capacity which is based on chemical precipitation, ion exchange and crystal growth (De Boodt, 1991). Five percent beringite mixed into the upper 30 cm of a disposal site substratum successfully allowed the rehabilitation of a non-ferrous waste disposal site. The substratum at the test site contained 18550 ppm Zn, 138 ppm Cd, 4000 ppm Pb and 1075 ppm Cu (in dry soils). The beringite fixed these metals thus limiting their translocation to the growth environment. This permitted revegetation and the healthy and dense growth of metal-tolerant species in an otherwise highly phytotoxic soil. The vegetation worked against lateral wind erosion of metal-contaminated particles. Although percolation increased, beringite reduced the amount of metals (e.g., Cd and Pb) in the percolate by more than 85% (Vangronsveld et al., 1995).

In some cases the treatment of heavy metals-contaminated soils is by incineration which immobilizes most potentially toxic metals (Rulkens et al., 1995). The resulting slag then must be disposed of at a secure site. If volatile toxic metals such as As, Hg and Se are contaminants in a soil, there must be a chemical scrubber in the emissions system that captures them before they reach the atmosphere, are dispersed by winds and ultimately precipitate back to the earth's surface. Without a control on incinerator emissions, the metals can cycle back to pollute soils, from there to fluvial systems that receive erosion products from the soils, and to marine waters where rivers discharge into the sea.

Sludge

Sludge is a broad term that describes a category of solid wastes of sewage, agriculture, or industrial origin. Sludge may be loaded with potentially toxic metals so that its disposal or treatment before disposal in an ecosystem is important.

Sewage sludge has been disposed of by spreading it on land as an agricultural amendment. The nutrient components add to the fertilization of soils. However, if the sewage carries a heavy metals load, it may prove detrimental to foodcrops or to food animals that consume fodder and pass pollutants to populations higher in the foodweb. If heavy metals are mobile in a solute that moves through soils, surface water and groundwater would be at risk of contamination. Hah et al. (2000) examined the accumulation and mobility of Cu and Zn in soils that had been amended with poultry wastes over a 25 year period. These metals were added to poultry diets to promote weight increase and disease prevention. They found that the Cu and Zn accumulated close to the surface and had significantly higher concentrations in amended soils than in adjacent non-amended soils. The soil organic matter contained almost 47 % of the Cu and the easily reducible oxide contained about the same amount of the Zn. In both cases the metals are bioavailable and mobile.

When sewage sludge from population centers or animal/poultry wastes contains only small concentrations of potentially toxic metals, it may be certified for use as a soil amendment. However, certification is site specific in terms of metal mobilities from wastes to soils and from soils to surface and aquifer waters. It is also conditioned by whether foodcrops grown in amended soils will bioaccumulate the metals and present a risk over time to human or animal consumers. Municipalities may treat the sludge and remove potentially toxic metals making it safe and useable for land application. Without like treatment, industrial byproducts in sludge wastes disposed of on, or mixed with soils can add to heavy metals loading and thus present a threat to the life in the disposal environment. For example, the growth of the ciliate protozoan *Colpoda steinii* common in soils was reported to be strongly inhibited in soils amended with heavy metals-bearing sewage sludge compared to unamended control soils (Forge at al., 1993). Heavy metals toxicity impact on soil microorganisms and microbial processes and their effects on sustainability of productive agricultural soils was reviewed by Giller et al., (1998). There are also threats from hand-to-mouth transfer by children of metals in soils amended with metals-bearing sludge, from inhalation of dust stirred up from the soils, and as previously stated, from consumption of food-crops grown in the amended soils for humans and for domesticated food animals that can subsequently translocate through the foodweb to humans. Younas et al. (1998) reported that application of sewage sludge/waters loaded a large area of soils around Lahore, Pakistan, with Cd, Cu, Ni and Pb. Their health study found that the average blood levels of Cd and Ni were significantly higher in cancer patients than in control groups. This suggests a possible link between Cd and/or Ni ingestion and the type of cancer in the population.

In a best case scenario, heavy metals in waste water sludges may be recovered and sold. Asarco used mine discharge sludges as smelter feedstock. The recovery of saleable metals and the production of non-hazardous products made this an economically and environmentally beneficial process (Mosher, 1994).

Surface and Aquifer Waters

Remediation of waters presents special problems. Surface waters can be self-cleansing if the source of contaminants is anthropogenic (e.g., industrial effluents) and can be eliminated. If not, surface waters have to be put through a treatment facility before being distributed to consumers. A recent innovation is the use of nanofiltration in a full size water treatment plant. The installation captures anything down to 1/10,000th of the thickness of a human hair such as bacteria, viruses and pesticides, thus making chemical treatment unnecessary. At Val D'Oise, France, water from the Oise River decants for two days and is filtered through sand and charcoal before being pressured through spiralling tubes of nanofilters at 8-15 bars. Over three million ft^2 of filter surface is used to produce four million ft^3 of water that is now servicing 300,000 households. The cost of production is about 10 cents more than conventional treatment and the capital investment for the plant was $ 150,000 (Pigeot, 2000). Heavy metals in contaminated waters might be susceptible to capture by this method or a modification of it.

As discussed later in this chapter, acid mine drainage from abandoned underground mines or surface tailings is laden with heavy metal contaminants and discharges into streams, rivers and lakes. This is a global environmental problem without a definitive economically feasible short-term solution. Acid mine or acid industrial drainage captured in ponds is most often neutralized with slaked lime [Ca(OH)$_2$], soda ash (Na$_2$CO$_3$), or caustic soda (NaOH). Many heavy metals form oxides upon neutralization and precipitate for subsequent disposal or recycling. Sediments deposited from polluted waters may have served as sinks for the metals. Remediation methodologies for sediments are discussed in the next section.

Remediation of heavy metals-contaminated subsurface waters containing leachates from waste disposal sites follows an isolation/containment/pressure directed extraction/treatment sequence. This is illustrated in Figure 8-1. The extend of a contaminated subsurface area is determined using chemical analyses of water pumped from an aquifer. A physical barrier is then implaced around the isolated zone to prevent migration and dispersion of heavy metals contaminants along flowpaths. Injection wells moving fluids under pressure drive the contaminated waters to extraction wells from which they are piped to treatment facilities. Monitoring wells outside the physical containment barrier provide security that contaminants are not escaping.

Conceptually, metals-contaminated aquifer waters could be treated at a well head if they are to be used as potable waters. In West Bengal, for example, well waters are polluted with As as described in Chapter 1. The water can be made potable if it passes through purification media. Unfortunately, purification kits cost US $ 14 which exceeds a monthly wage for the farmers who draw on the well waters. Either treatment facilities have to be built or alternate sources of clean water have to be established from rivers or deeper aquifers.

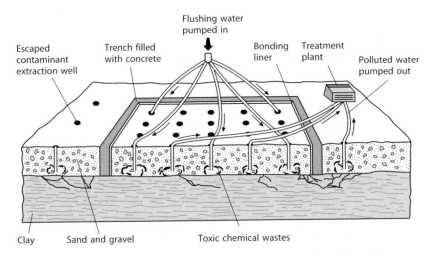

Fig. 8-1. General scheme for remediation of groundwater pollution from the Lipari Landfill Superfund site. A similar system was used at Love Canal, the first Superfund site. Chemical pollutants are contained and isolated by a concrete barrier. Removal of polluted waters is by pressurized flushing and pumping of the extract to a treatment plant. The security of groundwater is assured by monitoring wells outside the barrier (modified from Turk and Thompson, 1995)

Bioremediation in Soil, Sludge and Water Reclamation

The use of plants to clean soils contaminated with potentially toxic metals and make them safe and useable for humans is designated in different publictions as phytoremediation, bioremediation, botanical bioremediation, and green remediation (Chaney et al., 1997). Phytoremediation of metals-contaminated media has three tracks.

The first is phytoextraction whereby a plant accumulates metals in its shoots, is harvested, field dried, and ashed to recover valuable metals, or safely disposed of. Phytoremediation will be discussed in some detail after comments on phytostabilization and phytovolatilization. Phytostabilization is a process through which metals are not removed from polluted soil but instead are immobilized *in situ* as stable non-bioavailable forms or transformed to non-toxic species. This may be through the development of transgenic plants with increased metal binding capacity to reduce metal translocation from roots to edible parts and thus prevent access to a food chain or a non-food crop (e.g., Cd to rice or tobacco). It may be through soil amendments that induce the precipitation of toxic metals to insoluble compounds at roots such as Pb^{2+} to chloropyromorphite by the addition of phosphate (presently used). It may be by amendments now being investigated that could transform toxic metal species to non-toxic species such as Cr^{6+} which can be highly toxic to plants and soil organisms, to the highly insoluble Cr^{3+} (in research). Phytovolatilization is the third track. Research into plants that transform species of Hg, Se and perhaps As to forms which enter a vapour phase (e.g., Hg^0 or dimethylselenide) is giving encouraging results. Rugh et al. (1996) reported that there was a successful transference of microbial mercuric reduction genes from bacteria (microbes) to higher plants (e.g., *Arabidopsis thaliana* and other species). These transgenic plants reduce soil mercuric ions to Hg^0 which can volatilize from contaminated soils. The result is that there is a lesser risk that the very toxic methylated Hg will form in soils. A concern is the emission of Hg^0. However, the amount of Hg^0 added to the atmospheric load is negligible when compared to additions from volcanoes or from anthropogenic sources (e.g., combustion of fossil fuels for electricity generation). In the case of Se, Terry and Zayed (1994) found that both plants and soil microbes stimulated biosynthesis and volatilization of dimethyl selenide.

Of the three approaches to bioremediation, phytoextraction is close to being a viable economic option for the cleanup of metals-contaminated soils versus excavation, treatment, disposal, and replacement with clean soil, or flush and wash techniques. It is an outgrowth of principles and processes that have been used in biogeochemistry applied to mineral prospecting (Brooks, 1972, 1983; Kovalevskii, 1979, 1991; Brooks et al., 1995). Some plants will exclude metals from uptake. Other will absorb metals in amounts proportional to their

Table 8-1. Example of plant species which hyperaccumulate over 1% of a potentially toxic metal in their shoot dry matter in field collected samples (about 100 times higher than concentrations tolerated by normal crop plants). Cadmium maximum is included for comparison (after Chaney et al., 2000)

Metal	Species	Maximum Level in Leaves, ppm	Location
Zn	*Thlaspi calaminare*	39,600	Germany
Cu	*Aoellanthus biformifolius*	13,700	Zaire
Ni	*Phyllanthus serpentinus*	38,100	New Caledonia
Co	*Haumaniastrum roberti*	10,200	Zaire
Se	*Astragalus racemosus*	14,900	Wyoming
Mn	*Alyxia rubricaulis*	11,500	New Caledonia
Cd	*Thlaspi caerulescens*	1,800	Pennsylvania

concentrations in soils. The plants that will bioaccumulate metals to contents in excess of those of the soils are the subjects of phytoextraction research. To be useful in phytoextraction, an accumulator plant must be able to tolerate high metal concentrations without suffering phytotoxic effects. Plants that uptake bioavailable metals to concentrations of 100 to 1000 ppm are called accumulator species and those that uptake > 1000 ppm are called hyperaccumulators (Brooks et al., 1979). Table 8-1 gives examples of plant species which hyperaccumulate from more than 1% to nearly 4% of a potentially toxic metal in their shoot dry matter.

Cunningham and Berti (1993) calculated that phytoextraction is economically feasible if plants can concentrate about one percent of a metal (dry weight) in their shoots when grown in their natural soils. They gave a cost of $8 – $24 million for every meter of contaminated soil removed per hectare (2$^1/_2$ acres) which included disposal in a hazardous waste landfill and replacement of the excavated mass with clean soil. Moffat (1995) estimated a cost of $750,000 – $3,750,000 for excavation of a hectare to a depth of 1 m. The disparity in the costs may be that Moffat's figures did not account for varying soil conditions or include the disposal and soil replacement. However, in comparison with conventional excavation, incineration, and pump and treat systems for remediation of heavy metal-contaminated soils, phytoremediation can be only about 20% as costly. Thus, the efficiency and cost effectiveness parameters for each cleanup site have to be worked out before one or a combination of remediation technologies are applied to a problem.

Phytoextraction focuses on the hyperaccumulator plants. Many of these species have been identified especially from research on their potential use for mineral prospecting. For example, the species *Alyssum Linnaeus (Cruciferae)* is a hyperaccumulator of Ni in New Zealand and New Calendonia (Brooks et al., 1979) as are *Alyssum bertolonii* in Italy (Anderson et al., 1999), *Berkheya coddii* in South Africa (Robinson et al., 1999) and *Phyllanthus serpentinus* in New Caledonia (Kersten et al., 1979). *Haumaniastrum roberti* is a hyperaccumulator of Co in Zaire (Brooks, 1977), *Astragalus racemosus* of Se in Wyoming (Beath et al., 1937), and *Thlaspi caerulescens* of Zn and Cd in the Mediterranean area (Knight et al., 1997; Escarre et al., 2000). *Thlaspi calaminare* is a Zn hyperaccumulator in Germany (Reeves and Brooks, 1983), and *Iberis intermedia (Brassicaceae)* and *Biscutella laevigata* are hyperaccumulators of Tl in southern France (Anderson et al., 1999; Leblanc et al., 1999).

In general, hyperaccumulator species such as *Thlaspi caerulescens* have low biomass production, a short taproot which reaches a small volume of soil, and are small so that they cannot be mechanically harvested (Ernst, 1996). If these properties can be improved and difficulty in mechanical harvesting overcome, their utility for phytoremediation of heavily polluted soils would be enhanced. Manual harvesting in developing nations where employment opportunities are limited and populations high is a possibilty. Another potential problem with the use of hyperaccumulator plants is that of organisms feeding on them and being harmed. Chaney et al. (1997) report that high metal contents (e.g., Ni in leaves) may reduce herbivory by chewing insects and reduce the incidence of bacterial and fungal diseases. Further, sheep, goats and cattle avoid *Thlaspi* (Zn and Cd) and *Alyssum* (Ni) hyperaccumulators so that the risk to animals may not be significant (Chaney et al., 2000). Because phytoextraction is a long-term process, it should not be considered in place of excavation when short-term remediation is necessary to limit exposure and health risks (e.g., to young children from contact with Pb-polluted soils).

An ideal hyperaccumulator plant for use in bioremediation by phytoextraction should be indigenous to a climate, have roots that reach a reasonably large volume of soil, grow quickly, have a high rate of transfer of a target metal from roots to shoots, have a high shoot biomass and a good density of growth. Where one or more of these desired properties are wanting, it may be possible to improve them by good agricultural management such as weeding or using different fertilizer treatments. This was done for *Alyssum bertolonii*, a Ni hyper-accumulator cited above (Anderson et al., 1999). The biomass was increased from 4.5 tons/acre to 12 tons/acre without a significant loss of the Ni concentrations (7600 mg/kg). Chaney et al (1997) calculated that a biomass yield of about 20 tons/acre and contents of 10–40% of a target metal (Zn, Cu, Ni, Co) is required to recycle the metals at a profit using standard metallurgical methods. For many heavy metals, soil amendments such as the introduction of EDTA or acidification increases metal solubility, mobility to phytoextractor vegetation and rate of uptake and translocation from roots to shoots. For example, soils contaminated by a Zn smelter that operated for 80 years near Palmerton, Pennsylvania contained up to 10000 ppm Zn (1%) and 100 ppm Cd (dry weight). Li et al. (1997) lowered soil pH to 6.7 using nitrate salts. This favored the uptake and accumulation of both metals in *Thlaspi caerulescens* shoots compared to the species grown in control plots. A second harvest gave double the contents to about 20g/kg Zn and 200 mg/kg Cd in shoots (dry weight). Lettuce grown adjacent to *Thlaspi caerulescens* showed little increase in uptake demonstrating the influence of a crop vegetable on bioaccumulation of metals.

Otherwise, indigenous hyperaccumulator plants can be domesticated and bred to improve cultivar properties or they may be genetically engineered to improve critical properties such as metal tolerance, rate of metal uptake, amount of accumulation and increased biomass (Karenlampi et al., 2000). Among the most important factors that influence bioremediation are soil pH, moisture, bioavailability and chemical characteristics of a metal (or metal species), moisture, temperature, aeration or redox potential, nutrient supply, and soil texture. Foodcrops grown in soils containing heavy metals should naturally discriminate against the cumulative uptake of certain metals. If not, this property may be imparted to a foodcrop through genetic engineering. Brooks (1998) edited a volume that identifies plants that hyperaccumulate heavy metals and discusses their role in phytoremediation and other areas of scientific interest (e.g., phytomining, archaeology). Chaney et al. (1997, 1999, 2000) have given comprehensive assessments of phytoremediation of metals-contaminated soils including the benefits and limitations and in the 1999 paper give a 7 step procedure to develop commercial phytoremediation technology (Table 8-2). Their papers carry extensive up to date bibliographies and are recommended reading for those interested in bioremediation.

In addition to using vegetation hyperaccumulators to remediate heavy metals-contaminated soils, aquatic vegetation can be useful to cleanse polluted waters. This type of bioremediation is called rhizofiltration remediation. It can be used in areas where open drains or canals normally carry contaminated industrial and sewage effluents for discharge into lake, river, lagoon, sea or ocean waters, harming ecosystems. To deal with this problem, environmental geoscientists design managed, engineered wetlands through which contaminated waters and their entrained sediment load are channeled for remediation. Figure 8-2 gives a plan for a managed, engineered wetlands that was proposed for Manzalah lagoon, Nile delta. As the waste water and sediment load moves through the first stage of the wetlands, submergent vegetation serves to slow water flow and promote settling.

Table 8-2. Stages to develop commercial phytoremediation technology via domestication and breeding of improved hyperaccumulator plant species. Adapted from detailed text in Chaney et al., (1999)

1. Select hyperaccumulator plant species that lend themselves to domestication for use in a commercial phytoextraction system.

2. Collect seeds from plants in their natural ecosystems, bioassay their potential for use in phytoextraction, and breed improved cultivars for commercial operations.

3. Identify agricultural management practices to improve biomass, and amount and rate of extraction.

4. Develop crop management practices to improve plant density and effectiveness of phytoextraction.

5. Use biomass to produce energy and recover metal-rich ash in forms that can be disposed of safely at existing costs or use ash as ore for economic benefit. Combustion system must have up-to-date emission controls to capture particulate and volatilized metals.

6. Develop commercial operations for production of metal-rich biomass for recovery of energy and metals.

7. Identify and avoid negative or adverse aspects of phytoremediation by use of hyperaccumulator plants where best for society (e.g., rapid removal of Pb-loaded soils from populated areas) or predatory ecosystem inhabitants.

In the second stage, cattails, rushes and other emergent aquatic vegetation sequester and retain heavy metals. This is followed by interaction with rooted submerged forms that further remove heavy metals during the third phase of the remediation process. Stage 4 is perhaps the most critical as the final stage in the water-cleansing sequence. Here floating vegetation (water hyacinths) absorbs several potentially toxic metals (e.g., As, Cd, Hg, Pb) from the waters to the degree that the cleansed water can provide the environment for controlled fishery development until fingerlings can be released into the lagoon eco-

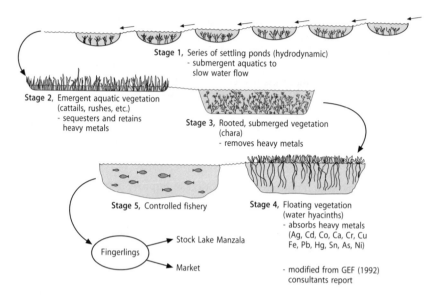

Fig. 8-2. Plan for managed, engineered wetlands for Manzalah lagoon, Nile delta, Egypt (modified by Siegel, 1995, after GEF, 1992)

system. Water hyacinths are efficient accumulators of several metals (Table 8-3) and have the requirements for being used as remediator plants described in the next paragraph. Atkinson et al. (1998) reported that biosorption with any type of biomass to passively sorb and immobiize solubilized heavy metals in a waste stream at a metal-plating plant in South Africa worked well but at the time of the research was not as cost effective as chemical precipitation.

Growing and harvesting remediator plants over time will reduce the heavy metal load of soils to natural levels so they no longer present an environmental threat. It can also be used for industrial biomass energy if volatilized metals (e. g., Hg and As) are captured before emission but certainly not for domestic combustion. The resulting ash may be recycled as an "ore" of one or more than one metal. Because of the heavy metal content, the harvested biomass is not useful as an animal feed or as a fertilizer since these uses would recycle the metal contaminants to the foodweb (Siegel, 1995). If the harvested plants are to

Table 8-3. Pollutant removal potential for water hyacinths (modified from Stephenson et al., 1980). NG = Not given

Metal	Accumulation g/ton dry wt.	Value if Recoverable ($ US)	
Ag	650	4.46/troy oz.	2/23/2001
Cd	670	3.80/lb	12/19/2000
Co	568	14.50/lb	2/23/2001
Cr	286	0.50/lb	1998
Cu	570	0.80/lb	2/23/2001
Fe	70	0.06/lb	2/23/2001
Hg*	136	2.00/lb	2/26/2001
Ni*	454	2.98/lb	2/23/2001
Pb	3200	0.30/lb	2/23/2001
As	NG	0.40/lb	6/16/2000
Sn	NG	2.33/lb	2/23/2001
Zn	NG	0.51/lb	2/26/2001

* Accumulation data from Wolverton et al. (1976).

be disposed of or if there are toxic metals residues from energy generation or ore-winning processes, a secure disposal site for them must be suitably prepared and monitored. The growing and harvesting of remediator plants has not been used in large scale projects but results from experimental plots are most encouraging. For example, Vetevier grass was used for phytoremediation of soil contaminated with heavy metal (Chen et al., 2000). The grass accumulated 218 g/ha of Cd from a soil with 0.33 mg/kg Cd. In California, the Ni hyperaccumulator *Streptanthus polygaloides* removed 100 kg of Ni per hectare (Anderson et al., 1999).

Microflora/bacteria have an important role in remediation of heavy metals contaminated soils. They may be introduced into polluted soils to alter ambient redox conditions. This mobilizes potentially toxic metals from sites on soil particles into fluids moving under pressure to collection or well recovery sites. For example, a change from reducing to less reducing or oxiding conditions can liberate metals such as Cd, Co, Cu, Ni and Zn to the soil water environment whereas a change from oxidizing or slightly reducing to stronger reducing environments can do the same for elements such as As, Fe, Mn and Mo. This is called bioleaching. The mobilized elements may then be captured by chelating agents/extractants such as EDTA or citric acid and transported until captured at a recovery well or collection site for treatment, recycling, or disposal. For example, in experimental projects, interaction of rainwater sludge with *Thiobacillus ferrooxidans*, an iron oxidizing strain, resulted in total extraction of Cd, Co, Cu and Ni from the sludge (Gomez and Bosecker, 1999). The strain *Thiobacillus thiooxidans*, a sulfur-oxidizing strain mobilized 80% of Cd, Co, Cu and Zn from the sludge. In both experiments Pb was not detected in the leachate. In other cases, heavy metals were extracted from oils and wastes with citric acid. They formed citrate complexes which were readily biodegraded by *Pseudomonas fluorescens* for capture and disposal (Francis and Dodge, 1998). Rittle et al., (1995) performed laboratory experiments in which they were able to effect bacterial sulfate reduction to induce precipitation of As^{3+} and immobilize it in sediments. They proposed that this may be a process by which other heavy metals in sediments can be immobilized.

Sediment (Fluvial, Lacustrine, Estuarine, Marine)

Metal contaminants in sediments in ponds (e.g., industrial and mining wastes), lakes, streams and rivers, estuaries, inland seas, oceans and associated wetlands present special problems. There are flowing waters, currents and overlying waters to deal with and the metal contamination may be widespread and diffuse. An environmental geochemist has to know the factors that can mobilize immobile metals in sediments to decide on remediation possibilities. These include lowering of the pH, changing reduction/oxidation conditions, inorganic and organic complexation, and microbial mediation. The metals are assessed as well for their response to barriers to mobility such as adsorption, absorption, controlled sedimentation and chemical precipitation. With these factors understood and known for metals at specific sites, the best way to deal with contaminated sediments can be evaluated (Förstner, 1996; Schuiling, 1990).

Dredging is discouraged as a heavy metals-contaminated sediment cleanup option if it brings about changes in environmental conditions that lead to mobilization and bioavailability. This is especially important if dredging exposes reduced sediment to oxidizing conditions. Table 2-4 quantifies the release of potentially toxic metals that has been documented after such an exposure. Similarly, in experimental work, the oxidative release of As from industrial solid wastes at a pH of 5 was complete in 5 weeks whereas the release of Zn began after 5 weeks and was continually enhanced (Förstner, 1996). In other experiments 20–90 % of pyrite-bound metals (e. g., As, Ni) were released and bioavailable in a day or less by exposure to oxidation conditions in sea water. The solution to dealing with metals-contaminated sediments immobile under existing environmental conditions is to leave them "stored" and undisturbed in the anoxic system. Reduced sediments loaded with potentially toxic metals (from industrial wastes) in areas susceptible to changing conditions such as harbors, bays or specific aquatic disposal sites can be further protected by capping them with clean sediment to better isolate them from the overlying oxic waters (Förstner, 1996).

Underground Mine Workings in Sulfide-Bearing Rock, Surface Mine Spoils

Acid mine drainage from the closed East Avoca and West Avoca mine workings discharges overall an average of 2.5 million litres daily into the Avoca river. This adds some 100–200 kg Zn, 10–15 kg Cu, 1–2 kg Pb and 1600 kg S to the river system daily (Galagher and O'Connor,

1996). The discharging waters are highly acid. Except for acid tolerant, metal-accumulating algae, aquatic life downstream from the discharge flow is non-existent. Given the amount of metal leachable from spoil heaps, backfill tailings, open pits and shafts, and the rate at which metals and sulfur as SO_2 are leaching from these materials, it could take 1000 years for natural cleansing to allow the Avoca river downflow of the acid mine discharge to be able to sustain aquatic life as does the river upstream of the discharge site. Herr and Gray (1997) studied Zn, Cu and Fe in the < 2 μm size fraction in surface to 30 mm riverine sediment layers below the Avoca mines. Cadmium was not present above a detection limit of 0.1 μg/g. The researchers concluded that pH regulates adsorption and desorption of the metals in sediments. The background value for Zn upstream of the mines' drainage is 84 ppm. This value decreases initially where the acid drainage enters the river waters. Downflow where there is complete mixing of the acid mine drainage with river water the Zn contents of the sediment rises to 681.3 ppm. Where the pH of river water rises to 8.7 downstream of a fertilizer plant about 7 km from the mines, the Zn concentration maximizes at 891 ppm. With rising pH, Zn precipitates and Cu co-precipitates with Fe. The metal concentrations are higher in floc material and surface ochre sediments compared to subsurface sediments. The scenario for the Avoca mines is repeated with lesser or greater intensity worldwide.

In a series of papers, hyperaccumulators were proposed for phyto-remediation/phytomining of heavy metals-bearing mine wastes tailings piles (Robinson et al., 1999; LeBlanc et al., 1999; Anderson et al., 1999). The bioaccumulated metals can be recovered sulfur-free and recycled if the economics works out (phytomining) or may be disposed of at a secure site isolated from ecosystems.

Remediation of heavy metals-bearing acid mine drainage could be accomplished in theory but is difficult, costly, and time-consuming. Acid mine drainage develops when the mineral pyrite, a significant component of sulfide ores is oxidized. The oxidation takes place as a bacterially mediated oxidation step reaction as follows:

$$2\ FeS_2 + 2\ H_2O + 7\ O_2 = 2\ Fe^{2+}_{(aq)} + 4\ SO_4^{2-} + 4\ H^+$$

$$2\ Fe^{2+}_{(aq)} + 1/2O_2 + 2H^+ = 2\ Fe^{3+}_{(aq)} + H_2O$$

$$2\ Fe^{3+}_{(aq)} + 6\ H_2O = 2\ Fe(OH)_3 + 6\ H^+$$

Summing the above reactions gives the origin of acid mine drainage:

$$4\ FeS_2 + 15O_2 + 14\ H_2O = 4\ Fe(OH)_3 + 8\ SO_4^{2-} + 16\ H^+$$

Thus, for every mole of FeS_2 oxidized, 4 moles of H^+ are released to the environment. This is one of the most acidic reactions in nature. If acid mine drainage is not remediated it can foul potable water sources, delay natural cleansing of streams, and as discussed in Chapter 1 (Figure 1-3), kill or displace fish and other aquatic life in impacted ecosystems.

Reactions of oxidation products of FeS_2 with other minerals in mine tailings add to the acid drainage problem. For example, the principal Zn ore mineral sphalerite reacts as follows:

$$2\ ZnS + 2\ Fe_2(SO_4)_3 + 2\ H_2O + 3\ O_2 = 2\ ZnSO_4 + 4\ FeSO_4 + 2\ H_2SO_4$$

Similarly, the main ore mineral of Cu, chalcopyrite reacts as

$$CuFeS_2 + 2\ Fe_2(SO_4)_3 + 2\ H_2O + 3O_2 = CuSO_4 + 5\ FeSO_4 + 2\ H_2SO_4$$

intensifying the acid drainage problem.

Where heavy metals-bearing acid mine drainage becomes part of the fluvial system, metal-tolerant and bioaccumulating algae can thrive. The algae do bring about some natural remediation by absorbing large concentrations of potentially toxic metals. However, the biomass of algae and the amount of metal absorbed is not enough to make a great difference in the total load.

The host rock and the amount, distribution, and type of mineralization in a mine are the principal factors that have to be assessed in evaluating the possibility of remediation of acid drainage from an abandoned or working operation. Additional factors such as climate, seepage access of rainwater or snowmelt to underground mines through roofrock fractures, the extent and interconnections of tunnels, the volume of water that enters and its exit pathways have to be determined. Once the impacts of the various factors for a proposed remediation or alleviation project have been determined, a critical question that must be answered is whether the tunnels (or mine tailings) can be deprived of contact with oxygen. Conceptually, mine adits/shafts can be sealed and filled with water. The amount of oxygen in the water is far less than in the atmosphere so that this limits the oxidation of pyrite and the generation of acid mine waters. Flooding has been successful in alleviating the acid drainage at some abandoned mines. However, the implacement of pressure bulkheads to maintain a flooded condition is very costly and carries a high risk of catastrophic failure. Waring et al. (1999) believe that a low cost and passive acid mine drainage prevention strategy would be to control the atmosphere in closed underground mines. An atmosphere stripped of its oxygen and largely nitrogen would be the key to this strategy. This can prevent the oxidation and maintain stability of sulfide minerals without affecting the flow of water out of a mine. If research demonstrates that this is feasible technically and economically, a mine could be easily reopened if the price of one or more commodities went up and made the mining operation profitable again.

The sealing of mines and mine tailings against rainwater or snowmelt seepage is possible borrowing a methodology developed for landslide mitigation. It has been successful in many landslide prone areas such as in Japan and Hong Kong. The shape of the surface area to be protected against infiltration by water is generated from computer analysis of the topography. Drainage systems are installed to divert any water that might enter a mine or tailings piles, thus limiting interaction with pyrite, the generation of acid drainage, and its transport away from a source. A metal frame is molded to the shape of the surface and rock-bolted to the hillside. Next a layer of the imper-

meable material gunite, a thick slurry of sand and cement is hydrauli-
cally sprayed in place on the frame sealing the surface. Exit tubes are
in the slope for any drainage that might access a slope.

For open pit mines, interior drainage is stored in ponds and treated
before allowing a release to an ecosystem. This is done by neutralizing
the acid and precipitating or sorbing heavy metals. These may be recovered and recycled or securely disposed of. Clean water is then
released to surface waters. This is operational procedure for acid
drainage from underground mines and from industrial sources as
well. Conditions in the acid drainage may change, however, so that
treatments will have to change as well. Table 8-4 proposes a plan for
continuous monitoring of entry and exit waters with automated
equipment (Fytas and Hadjigeorgiou, 1995). The monitored data are
used to continually adjust the chemical treatment. In this way, possible
environmental damage from infiltration of unspent chemicals into
groundwater or from their discharge into streams is prevented.

Integrated passive wetland treatment systems are being used
with metal-enriched, moderate to severely acidic drainage from the
Upper Blackfoot Mining Complex northeast of Missoula, Montana
that discharges into the upper Blackfoot River (Sanders et al., 1999).
Silver, Pb and Zn deposits made up much of the complex until they
were closed in the 1960s. Sanders et al. (1999) wrote that it will take
several years to reach full capacity in this high elevation environment
and that the treatment will last for several decades. The systems concentrate Zn and Fe and to a lesser extent Cu, Pb and other metals. This
is similar to a plan described in the Bioremediation section above for
the planned treatment of metals-bearing sewage wastes and industrial
effluents before discharge into Lake Manzalah, Egypt. The U. S. Bureau
of Mines (1992) described an economically beneficial wetlands operation at an abandoned coal mine near Coshocton, Ohio, that remediates
acid mine drainage and associated heavy metals. The construction
cost of US $ 60,000 for the three-celled wetland reduces chemical treat-
ment costs by at least US $ 20,000 annually for a less than three year
payback time.

Table 8-4. Parameters that may be measured in a continuous monitoring program for acid mine drainage and its controlled remediation (adapted from text, Fytas and Hadjigeorgiou, 1995)

A. Physico-chemical parameters
   1. pH
   2. redox potential
   3. total dissolved solids
   4. Specific conductance
   5. dissolved oxygen

B. Metal (cations plus anionic species) concentrations
   1. Iron (Fe)
   2. Copper (Cu)
   3. Lead (Pb)
   4. Zinc (Zn)
   5. Cadmium (Cd)
   6. Mercury (Hg)
   7. Arsenic (As)
   8. Sulfate (SO_4)

C. Gas concentrations
   1. Oxygen (O_2)
   2. Carbon dioxide (CO_2)
   3. Sulfur dioxide (SO_2)
   4. Hydrogen sulfide (H_2S)

D. Flow rate of acid mine drainage and hydrostatic pressure

E. Meteorological conditions
   1. Rainfall
   2. Temperature
   3. Sunlight
   4. Windspeed

Emissions

Emissions of potentially toxic metals from anthropogenic sources are as volatiles and particulates, and as mineral dusts. These are often accompanied by toxic gas (e.g., SO_2). The emissions have varying degrees of ecotoxicity. Their direct impact on ecosystem populations by respiration or through soils was described in earlier chapters. The indirect environmental impact on ecosystem populations is through acid rain which affects surface waters and mobilization of heavy metals from soils. In recent years, legislation in many countries required the installation of chemical scrubbers and electrical precipitators at industrial, mining, electricity generating and incineration facilities. The legislation allowed for closure of dangerous emission sources or costly fines that increases with non-compliance. This has discouraged a "polluter pays and pass cost on to consumer" attitude. The result has been a significant global decrease in toxic emissions of heavy metals, sulfur and other toxicants (Rasmussen, 1998). Unfortunately, and to the detriment of local environments, specific facilities that produce toxic emissions (especially electricity generating plants) were excluded from adherence to the new laws by environmentally ignorant, and/or protective politicians bowing to constituent pressure, industry lobbying, or reelection campaign financing concerns. The phasing out of the use of Pb as an anti-knock component in gasoline in most countries has greatly reduced Pb loading in ecosystems.

Legislation in the United States allots facilities a specific mass of emissions. If a facility does not use its emissions quota, the amount not used can be sold to another facility. This emissions trading solves no problem but just transfers emission loading of the atmosphere from one area to another. Emission trading is bad public policy. New legislation should be proposed which that puts constituent health and the environmental systems security as a prime objective. Industrial interests pushed by lobbying and campaign contribution sweetners

to influence (purchase) congressional votes can be negated by constituent voters who favor the election of politicians who put people/-
environmental interests first.

Remediation and Ecosystem Sustainabilty

There are many remediation techniques and technologies applicable
to specific environmental problems involving potentially toxic metals
at specific sites. Some may alleviate a problem to a tolerable status and
others may eliminate it. The choice of an integrated remediation plan
depends on several measureable physical, chemical and biological
factors at the heavy metals-polluted location and the environmental
media involved. The cost of the cleanup is important in decision-
making on the method(s) to be used and the degree of remediation
achieved. This may be a one-time investment over a short period of
time but more likely is a long-term economic commitment. The
Love Canal cleanup took more than a decade. Costs include capital
investment as well as funds for monitoring and maintenance of a
remediation system. Investment in cleanup of potentially toxic metals
impacted environments can give environmental and/or economic
benefits. It can make an environment suitable to again support an
ecosystem and perhaps attract tourism, or suitable for reuse in development programs such as housing as at Love Canal.

CHAPTER 9

Decision Making
for Environmental Sustainability

The essence of environmental sustainability lies with a population's accessibility to several basic biological, chemical and physical needs. These needs are clean air, clean water, sufficient nutrition, contained and isolated waste disposal, natural resources including wood and fiber for shelter and clothing, metals and non-metals for manufactured goods, sources of energy, and safe living and working space. In some global areas these needs are met but at the expense of creating unsustainable conditions (e.g., from deforestation, from over fishing, from over cropping without addition of nutrients to soil, and from mining without care to acid and metal-bearing drainage). In other areas, with forethought, planning and execution of directives, a sustainable ecosystem (e.g., with respect to forestry and fisheries) can be established. If attained, sustainability can meet the needs of the present population. Ideally this can be achieved without compromising the ability to meet the needs of future generations with larger populations.

The reality of the present global condition is that the needs of numerous populations in transnational regions, countries, provinces and states, and cities, towns and villages are not being met. Without changes in existing global patterns of prioritized investment, the future for many regions is bleak. Brady and Geets (1994) believe that to maintain or reestablish vital sustainable environments, human and governmental vision must change so that "the exploitation of natural

resources, the direction of investments, the orientation of technological development, and institutional change are all focused on enhanc-
ing current and future potential to meet human needs and aspirations". When followed, these guidelines have attained the optimal
goal of allowing an efficient and satisfactorily complete generation to
generation pass through of natural resources and environmental quality and their inherent social and economic benefits. Let us examine
where we are now with respect to establishing sustainable environments while meeting the needs of peoples worldwide by limiting the
impact of potentially toxic metals on ecosystems.

Populations

The earth's human population is presently at more than 6 billion. The
rate of annual increase for the world during 1995 was 2% but has been
falling and is now about 1.4%. This indicates that global population
could double to 12 billion in about 50 years unless the annual rate
continues to fall. Estimates from the United Nations Population
Bureau suggest that there will continue to be further reductions in
annual growth rates and that the Earth's human population will stabil-
ize in 2050 at about 9 billion inhabitants, at least one-half again as
many as we have today. At present we are not able to satisfy the basic
needs of at least $1/4$ of the human population. To satisfy these needs
and move foward with social and environmental progress requires
careful and directed planning in which the good will of "have nations"
abets a redistribution of basic needs and commodities together with
education as to the care needed to sustain living environments. Economic assistance from more affluent nations, international financial
organizations, and NGOs has to be used in an incorruptible manner.
Likewise, technical expertise has to be adapted to local conditions and
capabilities for effective and efficient environmental problem solving.

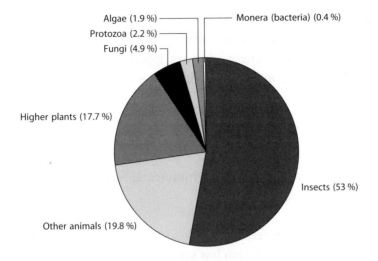

Fig. 9-1. Percentage distribution of identified living species by group. Virus comprise < 0.1% and do not show on the chart. Of the 19.8% of the Other Animals group, 2.9% are vertebrates Four to 20 times the number of life forms are estimated to exist (modified from Wilson, 1992)

National goals that can help the majority of inhabitants may mean that public policy changes have to be made. If these changes are followed and lead to sustainability of healthy productive environments, the purposes of environmental sciences research and recommendations that derive from it are justified.

Ecosystems house communities that are shared with humans by at least 1.4 million living species. Figure 9-1 illustrates the percentage distribution of living species by group. Insects dominate with about 53% of identified living species. Vertebrates and other animals comprise 19% and higher plants about 17% of living species, with the remaining identified forms distributed between fungi, protozoa, algae, bacteria (monera) and virus. Non-human populations are in competition with humans for sustenance and living space. Except for humans, the growth of these populations are self-regulated by eco-system limitations of accessibility to water, food and living space.

Humans have extended ecosystem limitations artificially in terms of access to clean water and by developing a green revolution that increases food stocks. These advances together with disease control, medical advances, improved nutrition plus better and more access to communication, education and research permit population growth.

Controlling Heavy Metals in Environmental Systems

The task faced by society to provide a sustainable environment secure from heavy metals pollution has several facets. The first is to identify environments in which anthropogenic loading of heavy metals puts ecosystems and their inhabitants at health risk. A second is to eliminate or alleviate anthropogenic sources of potentially toxic metal pollutants. The third facet is to remediate areas that have concentrations of heavy metals that present real threats to an ecosystem. A fourth is to protect populations from pollution while clean-up is proceeding. The fifth and sixth steps are directed towards identifying environments in which natural (baseline) concentrations and distributions of potentially toxic metals could present health problems and taking decisions to either avoid using these environments or use them for purposes that will not allow metals to gain access to life forms through water, food and air.

Toxicity to Sustainability of Living Environments

Water is the primary necessity to sustain life on earth. At present, about half the water available on earth from riverine, lake, and aquifer systems plus that from desalinization is being used. Yet one-quarter of the human population (1.5 billion people, mainly in lesser-developed nations) does not have access to safe drinking water, clean water for hygiene or water for sanitation. This causes the onset and spread of diseases that affect the ability of people to function and work efficently and to care for themselves and their families. About two million children die annually from bacterial, viral and parasitic diseases because of lack of safe water and sufficient water. The irregular geographic distribution of fresh water and lack of funds in most developing nations to import water and/or build distribution systems or waste water collection and treatment facilities makes this a continuing problem. A solution requires massive international subvention from nations and financial institutions with the technical expertise, economic where-with-all and moral sensitivity to help.

Toxicity sicknesses can impact populations from drinking and cooking with heavy metals-polluted water. They can also be caused by ingestion of food crops and/or food animals that incorporate and bioaccumulate toxic metals from polluted irrigation water, soils, or water bodies. They may originate with respiration. These sicknesses are not transmissible to others in a population. As such they are generally not as widespread as diseases that are readily transmissible and originate from biologically contaminated water and food, and from respiration or physical contact.

The damage to childrens' neurological functions and learning ability from the ingestion of Pb over time is well-documented worldwide. Recognition of the problem and identification of the principal sources of the Pb has resulted in global remediation that has given positive results. There are two principal sources and pathways by which Pb accessed afflicted children. One is through children in a home eating paint chips that contained Pb. Another is through children playing outside a house ingesting Pb from paint chips scraped off a house previous to repainting via hand to mouth contact with dirt and dust in a play area. In mining/smelting locations, smelter particulates become part of the earth's soil-dirt-dust surface and in these areas the incidence of Pb sickness in children from hand to mouth ingestion has been high. Respiration of smelter discharged Pb and gasoline emissions from leaded fuel is another pathway of this toxic element to the general population. The incidence of Pb sickness has been high downwind from smelters worldwide. The change to non-Pb-based paints and removal of Pb-bearing paints from homes, the use of unleaded gasoline in most countries, and enforced regulations on smelter emissions have greatly reduced the dangers of Pb to humans in ecosystem populations.

In Japan, waters containing cadmium (Cd) mobilized from mine wastes at an upland zinc (Zn) deposit in the 1950s discharged into the Jintsu River (Kobayashi and Hagino, 1965). Cadmium has a similar ionic size and the same ionic charge as does Zn and consequently has equal ability to form a bond. Geochemically, Cd will substitute for Zn in the mineral sphalerite (Zn[Cd]S) during the formation of a mineral deposit. Downstream in a lowland area, villagers used river waters contaminated with Cd draining from mine waste spoils for drinking, cooking (during which Cd become more concentrated), and for the irrigation of rice fields. The Cd bioaccumulated in the rice which is the staple in the villagers diet. During 5 to 7 years, consumers bioaccumulated Cd which interferred with the absorption of calcium (Ca) needed for replenishment in bone formation. The non-availability of Ca resulted in a slow, continuous loss of bone mass and development of cartiledge consistency matter in place of bone. Afflicted individuals were unable to move for the movement caused a pulling apart and breakage of the pseudo-bone matter and great pain. This was a type of osteomalacia which the Japanese called the itai-itai disease. Many villagers died as a result of the disease. This served as an event which signaled health professionals globally and people living in Zn mining areas of the possibilities and consequences of Cd pollution in drainage from mines and mine tailings. Care in disposal of mine wastes, clean-up of existing problem sites, and control of drainage so that there is no intrusion of surface or aquifer waters has been successful and the itai-itai disease caused by mining operations has been eliminated. The population can again use the river waters for drinking and cooking.

The Cd-contaminated soils present a problem and require that they be remediated before being used again for rice fields or that a rice strain be genetically engineered to discriminate against the accumulation of Cd. The former solution is preferred because the soil can be cleansed. If it is desired to change to crops that discriminate against Cd, there will be no concern about Cd uptake and accumulation. Cadmium remediation of the soils can be accomplished following the chemical extraction technique used by Urlings (1990) in Holland decribed in the previous chapter.

Wells yielding drinking water tainted with arsenic (As) have been responsible for deaths at limited localities in Argentina, China and Taiwan and for sickness elsewhere. On a large scale, however, as described in Chapter 2, poisoning from drinking and cooking with well water contaminated with As has afflicted at least 200,000 people in West Bengal, India and Bangladesh. This poses a real threat to the health of another several million people in both countries if the water can not be cleansed of As at the well head or if other sources of safe water can not be made available.

As described in Chapter 7, CH_3Hg^+ poisoning from the ingestion of tainted fish from Minamata Bay, Japan, during the 1950s killed 43 people, and permanently incapacitated 116 with permanent damage to their nervous systems. The fish bioaccumulated the toxic Hg species through the food web after the transformation of Hg^{2+} from untreated industrial discharge. After more than 35 years, the company that discharged the Hg accepted responsibility, provided compensation for the families of those who died, and created a fund for incapacitated victims and those who might suffer in the future from genetic changes. As a result of this case, more attention has been given to the possibility of Hg poisoning from the ingestion of foodfish that do have greater than 0.5 ppm Hg in their edible tissue. For example, mature swordfish, large tuna and shark contain greater than maximum permissible concentrations of Hg as methyl Hg. This Hg is undoubtedly a combination of natural loading and human activities loading (e.g., from combustion of coal, incineration of industrial waste). Pregnant women in the United States have been advised by a commission of the National Academy of Sciences to limit their intake of foodfish such as tuna and swordfish to 7 ounces, once a week lest toxic concentrations of methyl Hg bioaccumulate and pass through to a fetus putting a baby's future at risk.

Decision Making on Remediation, Alleviation, Prevention

Governmental economic resources for clean-up of environments polluted by potentially toxic metals that can be mobilized into a food web are in competition with other societal needs. Funds are limited but available to put where they will do the most good for the most distressed populations as well as for populations as a whole. Heavy metals pollutant generating sectors have been driven by a profit motive and only in recent years are they seeing economic and public relations benefits to using funds for efficient and effective pollution control systems. Investments in remediation or pollution control systems is likely not the result of "big industry seeing the light" or having a moral awakening. In some cases this has developed from harsh fines with the promise of more for the future. In other cases it has arisen from international lending institutions' requirements in order to obtain funding for development projects.

Available funding may not be enough to prevent, remediate or alleviate all the environmental problems that have originated from a heavy metals generating operation. Decisions makers can allocate funds following a rank system that sets priorities on the basis of which problems puts the ecosystem and its inhabitants at greatest immediate risk.

Comparative Risk Assessment

Comparative risk assessment (CRA) proposed by Davies (1996) and further considered by Ijjasz and Tlaiye (2000) evaluates environmental problems and "fix-it" choices that affect ecosystems and human health. Teams of scientists and environmental engineers work together with government, business, environmental groups and the public to do the ranking based on three principal norms. The first is to identify the most severe and threatening environmental problems from pollutants and/or pollutant sources for specific sites or broad areas. A second is to categorize the risks. The third is to develop priorities for clean-up, control and prevention strategies.

Three risk categories are used and appropriate to assess impacts of potentially toxic metals that concentrate in an environment. The first is health and may include the risk of a specific disease or health condition caused by exposure to one or more heavy metals. The health risk encompasses general effects of ingested heavy metals on reproductive, neurological, developmental and immunological processes. Identified problems can be combined for ranking purposes. A second is on ecological risks from heavy metals to all links in a food web from a single species to multiple ingestors that damage the integrity of an entire ecosystem. The third relates to the quality of life with respect to non-health factors such as the impact of heavy metals contamination of environmental media on recreation, commerce, property values and aesthetic values. The risks from exposure to pollutant metals are diverse, often affect different age groups (especially children) and may be fatal, permanently degrading a population's health status. Diseases caused by exposure to pollutant metals can also be short-term and reversible if diagnosed and treated in time.

After the risks are ranked, comparisions are made on the basis of how many people are exposed to individual risks and the geographical concentration of the risk factor(s). In some cases such as exposure to Pb and its neurological effects, chidren are considered as a separate population. The results are used to set regulatory priorities, perform benefit/cost analyses, target enforcement activities, set priorities for the allocation and strategic use of funds, and set priorities for action. This approach pays heed to knowledgeable community action committees and focuses environmental accountability thereby eliminating wasteful and inefficient fragmentation of responsibilities.

The process to calculate comparative risk assessments and evaluate them with respect to societal needs and values may take a year or two to complete. An approximate determination of the magnitude of a problem associated with pollution impacts and sources can be generated in about six months and is useful to nations in development which require a general focus for their assessment teams. Ijjasz and Tlaiye (2000) cite a case for which camparative risk assessment was useful to rank problems from environmental contamination associated with mining in Bolivia and come to decisions of remediation priorities. Mine sites were compared as to the risks they posed to people from heavy metals, to economic infrastructure, and to ecosystems through heavy metals contamination, acid drainage, and physical hazards. The comparative risk assessment guided decision makers in selecting mine sites to clean up first, how much to clean them up, and how to clean them up by identifying cost effective measures to the problems that arose from contamination. For example, dusty roads were paved, playgrounds built on mine wastes were sealed, and water supply pipes corroded by acid drainage were repaired.

Pre-Development Planning Against Heavy Metals Discharge

Comparative risk assessment provides the guidance necessary to prioritize existing environmental problems, evaluate technical methodologies to eliminate or alleviate them, and allocate funding for their remediation. A similar approach can be applied to preventing or limiting heavy metals release and ecosystem intrusion by pre-development planning. The contents of potentially toxic metals and other toxic chemicals in effluents or emissions that can be expected from a specific agricultural, mining or industrial project are known. Controls for each project and enforcement of regulations can prevent toxic metals intrusions of an ecosystem. The physical, chemical and biological pathways followed by one or a combination of these toxic components released to an ecosystem can be accurately predicted from studies of topography, surface and subsurface (aquifer) water and atmospheric current flow systems, soils and vegetation. With this knowledge, pre-development planning against environmental intrusion by heavy metals or other toxic chemicals can be done. This includes treatment and control systems to limit their discharge or emission to regulatory standards. These systems should be easily adapted for modulation if regulatory standards should change and new, more efficient and effective methodologies become available. Accidents can happen so that pre-planning can have automated shut down systems in place against effluent discharge and/or emissions, or back up systems ready to take on and clean effluent discharge or emission flow.

References

Al-Aasm, I.S., Clarke, J.D. and Fryer, B.J., 1998. Stable isotopes and heavy metal distribution in *Dreissena polymorpha* (zebra mussels) from western basin of Lake Erie, Canada. Envir. Geol., 33: 122–129.

Alloway, B.J., 1995. Soil processes and behaviour of metals. In (B.J. Alloway, Ed.), Heavy Metals in Soils, 2nd Ed., Blackie Academic & Professional, London, pp. 11–37.

AMAP, 1997. Arctic Pollution Issues: A State of the Arctic Environment Report. Arctic Monitoring and Assessment Programme, Oslo, 188 p.

Anderson, C.W.N., LaCoste, C.J., Leblanc, M., Robinson, B.H., Simcock, R., Stewart, R.B., Brooks, R.R. and Chiarucci, A., 1999. Phytomining for nickel, thallium and gold. J. Geochem. Explor., 67: 407–415.

Astrom, M., 1998. Mobility of Al, Co, Cr, Cu, Fe, Mn, Ni and V in sulfide-bearing fine-grained sediments exposed to atmospheric O_2: An experimental study. Envir. Geol., 36: 219–226.

Atkinson, B.W., Bux, F. and Kasan, H.C., 1998. Considerations for application of biosorption technology to remediate metal-contaminated industrial effluents. Water (South Africa), 24: 129–135.

Bagla, P. and Kaiser, J., 1996. India's spreading health crisis draws global arsenic experts. Science, 274: 174–175.

Beath, O.A., Eppsom, H.F. and Gilbert, C.S., 1937. Selenium distribution in and seasonal variation of type vegetation occurring on seleniferous soils. J. Amer. Pharm. Assoc., 26: 394–405.

Becker, P.R., 1993. Characterization of the arctic environment: marine biological resources. In (B.F. Molina and K.B. Taylor, Eds.), Workshop on Arctic Contamination. Arctic Res. of the U.S., 8: 69–76.

Berner, E.K. and Berner, R.A., 1996. Global Environment: Water, Air, and Geochemical Cycles. Prentice-Hall Inc., New Jersey, 376 p.

Bendell-Young, L.I., Dutton, M. and Pick, F.R., 1992. Contrasting two methods for determining trace metal partitioning in oxidized lake sediments. Biogeochemistry, 17: 205–219.

Bhattaacharya, B. and Sarkar, S.K., 1996. Total mercury content in marine organisms of the Hoogly Estuary, West Bengal, India. Chemosphere, 33: 147–158.

Bogle, E.W. and Nichol, I., 1981. Metal transfer, partition and fixation in drainage waters and sediments in carbonate terrain in southeastern Ontario. J. Geochem. Explor., 15: 405–422.

Bourg, A.C.M., 1995. Speciation of heavy metals in soils and groundwater and implications for their natural and provoked mobility. In (W. Salomons, U. Förstner and P. Mader, Eds.), Heavy Metals: Problems and Solutions. Springer, Berlin, pp. 19–31.

Brady, G.L. and Geets, P.C.F., 1994. Sustainable development: the challenge of implementation. Int'l Jour. Sustainable Development and World Ecol., 1: 189–197.

Breward, N. and Peachey, D., 1983. The development of a rapid scheme for the elucidation of the chemical speciation of elements in sediments. Sci. Total Environ., 29: 155–162.

Breward, N., Williams, M and Bradley, D., 1996. Comparison of alternative extraction methods for determining particulate metal fractionation in carbonate-rich Mediterranean soils. Applied Geochem., 11: 101–104.

Bromenshenk, J.J., Carlson, S.R., Simpson, J.C. and Thomas, J.M., 1985. Pollution monitoring of Puget Sound with honey bees. Science, 227: 632–634.

Brooks, R.R., 1972. Geobotany and Biogeochemistry in Mineral Exploration. Harper & Row, New York, 290 p.

Brooks, R.R., 1977. Copper and cobalt uptake by *Kaumaniastrum* species. Plant Soil, 48: 541–544.

Brooks, R.R., 1983. Biological Methods of Prospecting for Minerals. Wiley Interscience, New York, 322 p.

Brooks, R.R. (Ed.), 1998. Plants that hyperaccumulate heavy metals: their role in phytoremediation, microbiology, archaeology, mineral exploration and phytomining. CAB Int'l, 380 p.

Brooks, R.R., Dunn, C. and Hall, G.E.M., 1995. Biological systems in mineral exploration and processing. Ellis Horwood, New York, 538 p.

Brooks, R.R., Morrison, R.S., Reeves, R.D., Dudley, T.R. and Akman, Y., 1979. Hyperaccumulation of nickel by *Alyssum Linnaeus (Cruciferae)*, Proc. Royal Soc. London, B283, pp. 387–403.

Brothers, N.P. and Brown, M.J., 1987. The potential use of fairy prions *(Pachyptila turtur)* as monitors of heavy metal levels in Tasmanian waters. Marine Poll. Bull., 18: 132–134.

Brownlow, A.H., 1996. Geochemistry (2nd Ed.). Prentice Hall, New Jersey, 580 p.

Brundin, N.H., Ex, J.I. and Selinus, O.C., 1988. Biogeochemical studies of plants from stream banks in northern Sweden. J. Geochem. Explor., 27: 157–188.

Cardoso Fonseca, E. and Martin, H., 1986. The selective extraction of Pb and Zn in selected mineral and soil samples, application in mineral exploration (Portugal). J. Geochem. Explor., 26: 231–248.

Chaney, R.L., Brown, S.L., Li, Y-M., Angle, J.S., Stuczynski, T.I., Daniels, W.L., Henry, C.L., Siebielec, G., Makik, M., Ryan, J.A. and Compton, H., 2000. Progress in risk assessment for soil metals, and in-situ remediation and phytoextraction of metals from hazardous contaminated soils. Presented at USEPA Conference, Phytoremediation: State of the Science, 27 p.

Chaney, R.L., Li, Y.-M., Angle, A.J.M., Baker, R.D., Reeves, R.D., Brown, S.L., Homer, F.A., Makik, M. and Chin, M., 1999. Improving metal hyperaccumulator wild plants to develop commercial phytoextraction sytems: approaches and progress. In (N. Terry and G.S. Bunuelos, Eds.), Phytoremediation of Contaminated Soil and Water, CRC Press, Boca Raton, FL, 389 p.

Chaney, R.L., Malik, M., Li, Y.M., Brown, S.L., Brewer, E.D., Amgle, J.S. and Baker, A.J.M., 1997. Phytoremediation of soil metals. Current Opinions In Biotechnology, 8: 279–284.

Chao, T.T. and Theobald, P.K., 1976. The significance of iron and manganese oxides in geochemical exploration. Econ. Geol., 71: 1560–1569.

Chao, T.T., 1984. Use of partial dissolution techniques in exploration geochemistry. J. Geochem. Explor., 20: 101–135.

Chen, H.M., Shen, Z.G., Zheng, C.R. and Tu, C., 2000. Chemical methods and phytoremediation of soil contaminated with heavy metals. Chemosphere, 41: 229–234.

Chen, X., Yang, G., Chen J., Chen, X., Wem, Z. and Ge, K., 1980. Studies on the relation of selenium and Keshan disease. Bio. Trace Elem. Res., 2: 91–107.

Chester, R. and Hughes, M.J., 1967. A chemical technique for separation of ferromanganese minerals, carbonate minerals and adsorbed trace elements from pelagic sediments. Chem. Geol., 2: 249–262.

Christensen, J.W., 1991. Global Science. Kendall/Hunt Publ. Co., Dubuque, Iowa, 296 p.

Cleary, D., Thornton, I., Brown, N., Kazantis, G., Delves, T. and Worthington, S., 1994. Mercury in Brazil. Nature, 369: 613–614.

Connor, J.J. and Shacklette, H.T., 1975. Background geochemistry of some soils, plants, and vegetation in the conterminous United States. U.S. Geol. Survy. Prof. paper 574–F, 164 p.

Coquery, M., Cossa, D. and Martin, J.M., 1995. The distribution of dissolved and particulate mercury in three Siberian estuaries and adjacent Arctic coastal waters. Water, Air and Soil Poll., 80: 653–664.

Crounse, R.G., Pories, W.J., Bray, J.T. and Mauger, R.L., 1983. Geochemistry and man: health and disease; 1. essential elements, 2. elements possibly essential, those toxic and others. In (I. Thornton, Ed.), Applied Environmental Geochemistry, Academic Press, London, pp. 267-308; 309–333.

Cunningham, S.D. and Berti, W.R., 1993. Remediation of contaminated soil with green plants: An overview. In-Vitro Cell Dev. Biol., 29P: 207–212.

Darby, D.A., Adams, D.D. and Nivens, W.T., 1986. Early sediment changes and element mobilization in a man-made estuary marsh. In (P.G. Sly, Ed.), Sediment and Water Interactions (Springer, Berlin, pp. 343–351.

Darnley, A.G., Bjorklund, A., Bolviken, B., Gustavsson, N., Koval, P., Plant, J.A., Steenfelt, A., Tauchid, M. and Xie, X., with contributions by R.G. Garrett and G.E.M. Hall, 1995. A Global Geochemical Database: For Environmental and Resource Management. UNESCO, Earth Sciences 19.

Davies, J.C. (Ed.), 1995. Comparing Environmental Risks: Tools for Setting Government Priorities. Resources for the Future, Washington, D.C., 157 p.

Davis, J.C., 1986. Statistics and Data Analysis in Geology (2nd Ed.). John Wiley & Sons, New York, 646 p.

De Boodt, M.F., 1991. Application of the sorption theory to eliminate heavy metals from waste waters and contaminated soils. In (G.H. Bolt, M.F. de Boodt, M.H.B. Hayes and M.B. McBride, Eds.), Interactions at the Soil Colloid-Soil Solution Interface. NATO ASI Series, Series E: Applied Sciences, Vol. 190, Kluwer Academic Publishers, Dordrecht, pp. 293–320.

Degueldre, C., Triay, I., Kim, J.I., Vilks, P., Laaksoharuju, M. and Mielkeley, N., 2000. Groundwater colloidal properties: a global approach. Applied Geochem., 15: 1043–1051.

De Zuane, J., 1997. Handbook of Drinking Water Quality. John Wiley & Sons, Inc., New York, 575 p.

Dietz, R., Riget, F. and Johansen, P., 1996. Lead, cadmium, mercury and selenium in Greenland marine animals. Sci. Total Envir., 186: 67–93.

Dietz, R., Aarkrog, A., Johansen, P., Hansen, J.C., Riget, F. and Cleeman, M., 2000. Comparison of contaminants from different trophic levels and ecosystems. Sci. Total Envir., 245: 221–231.

Dipankar, D., Samanta, G., Mandal, B.K., Chowdhury, T.R., Chanda, C.R., Chowdhury, P.P., Basu, G.K. and Chakraborti, D., 1996. Arsenic in groundwater in six districts of West Bengal, India. Envir. Geochem. & Health, 18: 5–15.

Doelman, P., 1995. Microbiology of soil and sediments. In (W. Salomons and W.M. Stigliani, Eds.), Biogeodynamics of Pollutants in Soils and Sediments, Springer, Berlin, pp. 31–52.

Dunn, J., Russell, C. and Morrissey, A., 1999. Remediating historic mine sites in Colorado. Mining Engineer., 51: 32–35.

Drever, J.I., 1997. The Geochemistry of Natural Waters (3rd Ed.). Prentice Hall Inc., New Jersey, 436 p.

Ernst, W.H.O., 1996. Bioavailability of heavy metals and decontamination of soils by plants. Applied Geochem., 11: 163–167.

Escarre, J., Lepart, J., Riviere, Y., Delay, B., Lefebvre, C., Gruber, W. and Leblanc, M., 2000. Zinc and cadmium hyperaccumulation by *Thlaspi caerulescens* from metalliferous and nonmetalliferous sites in the Mediterranean area: Implication for phytoremediation. New Phytologist, 145: 429–437.

Fergusson, J.E., 1982 (reprinted with corrections, 1985). Inorganic Chemistry and the Earth: Chemical Resources, Their Extraction, Use and Environmental Impact. Pergamon Press, Sydney, 400 p.

Fergusson, J.E., 1990. The Heavy Elements: Chemistry, Environmental Impact and Health Effects. Pergamon Press, Oxford, England, 614 p.

Filipek, L.H. and Theobald, P.K., 1981. Sequential extaction techniques applied to a porphyry copper deposit in the basin and range province. J. Geochem. Explor., 14: 155–174.

Food and Agriculture Organization of the United Nations, 1991. Major Climatic Zones. FAO, Rome, 11 p.

Fordyce, F.M., Guangdi, Z., Green, K. and Xinping, L., 2000. Soil, grain and water chemistry in relation to human selenium-responsive diseases in Enshi District, China. Applied Geochem., 15: 117–132.

Forge, T.A., Berrow, M.L., Darbyshire, J.F. and Warren, A., 1993. Protozoan bioassays of soil amended with sewage sludge and heavy metals, using common soil ciliate *Colpoda Steinii*. Biology and Fertility of Soils, 117: 282–286.

Förstner, U., 1996. Waste treatment: geochemical engineering approach. In (Reuther, R., Ed.), Geochemical Approaches to Environmental Engineering of Metals. Springer, Berlin, pp. 155–182.

Francis, A.J. and Dodge, C.J., 1998. Remediation of soils and wastes contaminated with uranium and toxic metals. Envir. Sci. & Tech., 3: 3993–3998.

Fytas, K. and Hadjigeorgiou, J., 1995. An assessment of acid rock drainage continuous monitoring technology. Envir. Geol., 25: 36–42.

Gatehouse, S., Russell, D.W. and Van Moort, J.C., 1977. Sequential soil analysis in exploration geochemistry. J. Geochem. Explor., 8: 483–494.

Galagher, V. and O'Connor, P., 1996. Characterization of the Avoca mine site: geology, mining features, history and soil contamination study. Geol. Surv. Ireland, Tech. Rept MS96/4.

GEF (Global Environmental Facility Chairman's Report), 1992. United Nations Development Programme, United Nations Environmental Programme, World Bank, Washington, D.C.

Geostandards Newsletter, 1994. July Number, Vol. 18.

Giller, K.E., Witter, E. and McGrath, S.P., 1998. Toxicity of heavy metals to microorganisms and microbial processes in agricultural soils: A review. Soil Biology and Biochem., 30: 1389–1414.

Goldschmidt, V.M., 1937. The principles of distribution of chemical elements in minerals and rocks. J. Chem. Soc., pp. 655–673.

Gomez, C. and Bosecker, K., 1999. Leaching heavy metals from contaminated soil by using *Thiobacillus ferrooxidans and Thiobacillus thiooxidans*. Geomicrobiology J., 16: 233–244.

Gong, C. and Donahoe, R.J., 1997. An experimental study of heavy metal attenuation and mobility in sandy loam soils. Applied Geochem., 12: 243–254.

Gray, D.J., Wildman, J.E. and Longman, G.D. 1999. Selective and partial extraction analyses of transported overburden for gold exploration in the Yilgarn Craton, Western Australia. J. Geochem. Explor., 67: 51–66.

Hah, F.X., Kingery, W.L., Selim, H.M. and Gerard, P.D., 2000. Accumulation of heavy metals on a long-term poultry waste-amended soil. Soil Sci., 165: 260–268.

Hall, G.E.M. (Ed.), 1992. Geoanalysis. J. Geochem. Explor., 349 p.

Hall, G.E.M. and Bonham-Carter, G.F., 1998. Selective extractions. J. Geochem. Explor., 61, no. 1–3, 232 p.

Hall, G.E.M. and Pelchat, P., 1997. Comparison of two reagents, sodium pyrophosphate and sodium hydroxide in the extraction of labile metal organic complexes. Air, Water and Soil Poll., 99: 217–223.

Hall, G.E.M., Vaive, J.E., Beer, R. and Hoashi, M., 1996. Selective leaches revisited, with emphasis on the amorphous Fe oxyhydroxide phase extraction. J. Geochem. Explor., 56: 59–78.

Hasan, S.E., 1996. Geology and Hazardous Waste Management. Prentice Hall, New Jersey, 387 p.

Herr, C. and Gray, N.F., 1997. Metal contamination of riverine sediments below the Avoca mines, south east Ireland. Envir. Geochem. & Health, 19: 73–82.

Hoffman, S.J. and Fletcher, W.K., 1979. Selective sequential extraction of Cu, Zn, Fe, Mn, and Mo from soils and sediments. In (J.R. Watterson and P.K. Theobald, Eds.), Geochemical Exploration 1978. Assn. Explor. Geochemists, pp. 289–299.

Ilton, 1999. Chromium. In C.P. Marshall and R.W. Fairbridge (Eds.), Encyclopedia of Geochemistry, Kluwer Academic Publishers, Dordrecht, The Netherlands, pp. 81–82.

Ilton, 1999. Selenium. In C.P. Marshall and R.W. Fairbridge (Eds.), Encyclopedia of Geochemistry, Kluwer Academic Publishers, Dordrecht, The Netherlands, pp. 571–572.

Ijjasz, E. and Tlaiye, L., 2000. Comparative risk assessment. Poll. Management Discussion Note No. 2., World Bank, Washington, D.C., 4 pp.

Islam, M.R., Salminen, R. and Lahermo, P.W., 2000. Arsenic and other toxic elemental contamination of groundwater, surface water and soil in Bangladesh and its possible effects on human health. Envir. Geochem. & Health, 22: 33–53.

Jackman, A.P. and Powell, R.L., 1991. Hazardous Waste Treatment Technologies. Noyes Publications, New Jersey, 276 p.

Johnson, C.C., Ge, K., Green, K.A. and Liu, X., 2000. Selenium distribution in the local environment of selected villages of the Keshan Disease belt, Zhangjiakou District, Hebei Province, People's Republic of China. Applied Geochem., 15: 385–401.

Joiris, C.R., Moaternri, N.L. and Holsbeek, L., 1995. Mercury and polychlorinated biphenyls in suspended particulate matter of the European Arctic seas. Bull. Envir. Contam. and Toxicology, 55: 893–900.

Kabata-Pendias, A., 1992. Trace metals in soil of Poland – occurrence and behaviour. Trace Substances in Environmental Health, 25: 53–70.

Kabata-Pendias, A., 1993. Behavioural properties of trace metals in soils. Applied Geochem. Supplementary Issue No. 2, Environmental Geochemistry, pp. 3–9.

Kansanen, P.H. and Venetvaara, J., 1991. Comparison of biological collectors of airborne heavy metals near ferrochrome and steel works. Water, Air and Soil Poll., 60: 337–359.

Karenlampi, S., Verkleij, J.A.C., Van der Lelie, D., Mergeay, M., Tervahauta, A.I., Schat, H. and Vangronsveld, J., 2000. Genetic engineering in the improvement of plants for phytoremediation of metal polluted soils. Environmental. Poll., 107: 225–231.

Kersten, W.J., Brooks, R.R., Reeves, R.D. and Jaffre, T., 1979. Nickel uptake by New Caledonian species of *Phyllanthus*. Taxon, 28: 529–534.

Kido, T., Nogawa, K., Shaikh, Z.A., Kito, H. and Honda, R., 1991. Dose-response relationship between dietary cadmium intake and metallothioneinuria in a population from a cadmium-polluted area of Japan. Toxicology, 66: 271–278.

Kim, K-W. and Thornton, I., 1993. Influence of uraniferous black shales on cadmium, molybdenum and selenium in soils and crop plants in the Deog-Pyoung area of Korea. Envir. Geochem. and Health, 15: 119–133.

Klein, D.R. and Vlasova, T.J., 1992. Lichens, a unique forage resource threatened by air pollution. Rangifer, 12: 21–27.

Knight, B., Zhao, F.J., McGrath, S.P. and Shen, Z.G., 1997. Zinc and cadmium uptake by the hyperaccumulator *Thlaspi caerulescens* in contaminated soils and its effect on the concentration and chemical speciation of metals in soil solution. Plant and Soil, 197: 71–78.

Kobayashi, J. and Hagino, N., 1965. Strange osteomalacia by pollution from cadmium mining. Progress Rept. WP 00359, Okayama Univ., pp. 10–24.

Kovalevskii, A.L., 1979. Biogeochemical exploration for mineral deposits. Amerind Publ. Co. Pvt. Ltd. New Dehli, 136 p. Translation of a 1974 Russian edition.

Kovalevskii, A.L., 1991. Biogeokhimiia rastenii (Biogeochemistry of Plants). Nauka, Sibirskoe otdelenei, 288 p. In Russian.

Krumbein, W.C. and Graybill, F.A., 1965. An Introduction to Statistical Models in Geology. McGraw-Hill, New York, 475 p.

Lacatusu, R., Rauta, C., Carstea, S. and Ghelase, I., 1996. Soil-plant-man relationships in heavy-metal polluted areas in Romania. In: (R. Fuge, M. Billet, O. Selinus, Eds.) Environmental Geochemistry, 3rd Int'l Symp., Krakow, Poland, Applied Geochem., 11: 105–107.

Lacerda, D., Malm, O., Guimaraes, J.R.D., Salomons, W. and Wilken, R.-D., 1995. Mercury and the new goldrush in the south. In (W. Salomons and W.M. Stigliani, Eds.), Biogeodynamics of Pollutants in Soils and Sediments. Springer, Berlin, pp. 213–245.

Leblanc, M., Brooks, R.R., Petit, D., Deram, A. and Robinson, B.H., 1999. The phytomining and environmental significance of hyperaccumulation of thallium by *Iberis intermedia* from southern France. Econ. Geol., 94: 109–113.

Lefcourt, H., Ammann, E. and Eiger, S.M., 2000. Antipredatory behavior as an index of heavy-metal pollution? A test using snails and caddisflies. Archives Environmental Contam. and Toxicology, 38: 311–316.

Li, Y.-M., Chaney, R.L., Chen, K.Y., Kerschner, B.A., Angle, J.S. and Baker, A.J.M., 1997. Zinc and cadmium uptake of hyperaccumulator *Thlaspi caerulescens* and four turf grasses (abs.), Agron. p. 38.

Loring, D.H., Naes, K., Dahle, S., Matishov, G.G. and Illin, G., 1995. Arsenic, trace metals, and organic micro-contaminants in sediments from the Pechora Sea, Russia. Marine Geol., 128: 153–167.

Mackenzie, F.T., 1998 (2nd Ed.). Our Changing Planet. Prentice Hall Inc., New Jersey, 486 p.

Mason, B. and Moore, C.B., 1982. Principles of Geochemistry (4th Ed.). John Wiley & Sons, Inc., New York, 344 p.

Matschullat, J., Ellmingen, F., Agdemir, N., Cramer, S., Liebmann, W. and Niehoff, N., 1997. Overbank sediment profiles – evidence of early mining and smelting activities in the Harz mountains, Germany. Applied Geochem., 12: 105–114.

McLennan, S.M. and Murray, R.W., 1999. Geochemistry of sediments. In (C.P. Marshall and R.W. Fairbridge, Eds.), Encyclopedia of Geochemistry. Kluwer Academic Publishers, Dordrecht, The Netherlands, pp. 282–292.

McLennan, S.M. and Taylor, S.R., 1999. Earth's continental crust. In (C.P. Marshall and R.W. Fairbridge, Eds.), Encyclopedia of Geochemistry. Kluwer Academic Publishers, Dordrecht, The Netherlands, pp. 145–151.

Merian, E. (Ed.), 1991. Metals and Their Compounds in the Environment: Occurrence, Analysis and Biological Relevance. VCH, Weinheim, Germany, 1438 p.

Monsanto, 1995. Development of an integrated in-situ remediation technology. Technology Development Data Sheet.

Monteiro, L.R., Granadeiro, J.P. and Furness, R.W., 1998. Relationship between mercury levels and diet in Azores seabirds. Marine Ecology Progress Series, 166: 259–265.

Moffat, A.S., 1995. Plants proving their worth in toxic metal cleanup. Science, pp. 302–305.

Mosher, J., 1994. Heavy metal sludges as smelter feedstock: how Asarco cleans mine discharge water and scores economic benefits as well. Eng. & Mining Jour., 195: 25–30.

Nendza, M., Gies, A., Herbst, T. and Kussatz, C., 1997. Potential for secondary poisoning and biomagnification in marine organisms. Chemosphere, 35: 1875–1885.

Nickson, R.T., McArthur, J.M., Burgess, W.G., Ahmed, K.M., Ravenscroft, P. and Rahman, M., 1998. Arsenic poisoning of Bangladesh groundwater. Nature, 395: 338.

Nickson, R.T., McArthur, J.M., Ravenscroft, P., Burgess, W.G. and Ahmed, K.M., 2000. Mechanism of arsenic release to groundwater, Bangladesh and West Bengal. Applied Geochem., 15: 403–413.

Nordberg, G.F., Cai, S., Wang, Z., Zhuang, F., Wu, X., Jin, T., Kong, Q. and Ye, T. 1996. Biological monitoring of cadmium exposure and renal effects in a population group residing in a polluted area in China. Sci. Total Env., 1999: 111–114.

Norrgren, L., Petersson, U., Orn, S. and Bergqvist, P.-A., 2000. Environmental monitoring of the Kafue River, located in the Copperbelt, Zambia. Archives Environmental Contam. Toxicology, 38: 334–341.

O'Brien, D.J., Kaneene, J.B. and Poppenga, R.H., 1993. The use of mammals as sentinels for human exposure to toxic contaminants in the environment. Environmental Health Perspectives, 99: 351–368.

Pacyna, J.M., 1995. The origin of Arctic air pollutants: lessons learned and future research. Sci. Total Envir., 160/161: 39–53.

Pacyna, J.M. and Keeler, G.J., 1995. Sources of mercury in the Arctic. Water Air, and Soil Poll., 80: 621–632.

Perel'man, A. Geochemical barriers: theory and practical application. Applied Geochem., 1: 669–680.

Peters, R.W., 1999. Chelant extraction of heavy metals from contaminated soils. J. Hazardous Materials, 66: 151–210.

Pigeot, J., 2000. Le Figaro, June 20, p. 14.

Pirrone, N., Keeler, G.J. and Nriagu, J.O., 1996. Regional differences in worldwide emission of mercury to the atmosphere. Atmospheric Envir., 30: 2981–2987.

Ponchet, H., 2000. Le Point, July 7, p. 44.

Powell, M.I. and White, K.N., 1990. Heavy metal accumulation by barnacles and its implications for their use as biological monitors. Marine Envir. Res., 30: 91–118.

Rainbow, P.S., 1995. Biomonitoring of heavy metal availability in the marine environment. Marine. Poll. Bull., 31: 183–192.

Rasmussen, L., 1998. Effects of afforestation and deforestation on the deposition, cycling and leaching of elements. Agriculture, Ecosystems and Environment, 67: 153–159.

Reeves, R.D. and Brooks, R.R., 1983. European species of *Thlaspi L. (Cruciferae)* as indicators of nickel and zinc. J. Geochem. Explor., 18: 275–283.

Reuther, R. (Ed.), 1996. Geochemical Approaches to Environmental Engineering of Metals. Springer, Berlin, 221 p.

Richardson, D.H.S., 1992. Pollution monitoring with lichens. Richmond Publishing, Slough, SL2 3RS, Naturalists' Handbook, 76 p.

Rittle, K.A., Drever, J.I. and Colberg, P.J.S., 1995. Precipitation of arsenic during bacterial sulfate reduction. Geomicrobiology Jour., 13: 1–11.

Robinson, B.H., Brooks, R.R. and Clothier, B.E., 1999. Soil amendments affecting nickel and cobalt uptake by *Berkheya coddii*: potential use for phytomining and phytoremediation. Annals of Botany, 84: 689–694.

Rose, A.W. and Suhr, N.H., 1971. Major element content as a means of allowing for background variation in stream-sediment geochemical exploration. In (R.W. Boyle, Ed.), Geochemical Exploration. Can. Min. Metall. Spec. Vol. 11, pp. 587–593.

Rose, A.W., Hawkes, J.E. and Webb. J.S., 1979. Geochemistry in Mineral Exploration. Academic Press, London and New York, 657 p. 2nd Ed. of the book by Hawkes J.E. and Webb, J.S., 1962.

Rugh, C.L., Wilde, H.D., Stack, N.M., Thompson, D.M., Summers, A.O. and Meagher, R.B., 1996. Mercuric ion reduction and resistance in transgenic *Arabidopsis thaliana* plants expressing a modified bacterial merA gene. Proc. Natl. Acad. Sci., 93: 3182–3187.

Rulkens, W.H., Grotenhuis, J.T.C. and Tichy, R., 1995. Methods for cleaning contaminated soils and sediments. In (W. Salomons, U. Förstner and P. Mader,
Eds.), Heavy Metals: Problems and Solutions. Springer, Berlin, pp. 169–191.

Salminen, R. and Tarvainen, T., 1997. The problem of defining geochemical baselines. A case study of selected elements and geological materials in Finland.
J. Geochem. Explor., 60: 91–98.

Samiullah, Y., 1990. Biological monitoring of environmental contaminants:
animals. Univ. London, Monitoring and Research Centre, Tech. Rept. 37, 767 p.

Sanders, F., Rahe, J., Pastor, D. and Anderson, R., 1999. Wetlands treat mine runoff.
Civil Engin., 69: 53–55.

Savinova, T.N., Gabrielson, G.W. and Savinov, V.M., 1997. Trace elements in seabirds from the Barents Sea area, 1991–1993. The AMAP Int'l Symp. on Environmental Pollution in the Arctic. Extended Abstracts. Tromsø, Norway,
pp. 224–26.

Schuiling, R.D. 1990. Geochemical engineering, some thoughts on a new research
field. Applied Geochem., 9: 553–559.

Scokart, P.O., Meeus-Verdinne, K. and De Borger, R., 1983. Mobility of heavy
metals in polluted soils near Zn smelters. Water, Air and Soil Poll., 20: 451–
463.

Selinus, O., 1995. Large-scale monitoring in environmental geochemistry. Applied
Geochem., 11: 2251–260.

Selinus, O., Frank, A, and Galgan, V., 1996. Biogeochemistry and metal biology. In
(Appleton, J.D., Ed.), Environmental geochemistry and health with special
reference to developing countries. Geol. Soc., Spec. Publ. 113, pp. 81–89.

Shacklette, H.T. and Boerngen, J.G., 1984. Element Concentrations in Soils and
Other Surficial Materials of Conterminous United States: An Account of
the Concentrations of 50 Chemical Elements in Samples of Soils and Other
Regoliths. U.S. Geol. Survey Prof. Paper 1270.

Siegel, F.R., 1985. In-situ recovery of suspended sediments from streams. Jour.
Geol. Education, 33: 132–133.

Siegel, F.R., 1990. Exploration for Mississippi-Valley type stratabound Zn ores
with stream suspensates and stream sediments, Virginia, U.S.A. J. Geochem.
Explor., 38: 265–283.

Siegel, F.R., 1992. Geoquímica Aplicada. Serie de Química, Monografía No. 35,
Organización de los Estados Americanos, Washington, D.C., 175 p.

Siegel, F.R., Slaboda, M.L. and Stanley, D.J., 1994. Metal loading in Manzalah
lagoon, Nile delta, Egypt: implications for aquaculture. Envir. Geol., 23: 89–98.

Siegel, F.R., 1995. Environmental geochemistry in development planning: an
example from the Nile delta, Egypt. J. Geochem. Explor., 55: 265–273.

Siegel, F.R. and Segura P., A., 1995. Biogeochemistry for future mineral resource
exploration programs in the Central America-Caribbean region. In (R.L.
Miller, G. Escalante, J.A. Reinemund and M.L. Bergin, Eds.), Energy and
Mineral Potential of the Central American-Caribbean Regions. Springer,
Berlin, Earth Science Series 16, pp. 329–334.

Siegel, F.R., 1996. Natural and Anthropogenic Hazards in Development Planning. Academic Press, New York and Landes Bioscience, Austin, 300 p.

Siegel, F.R., 1998. Geochemistry, Metal Toxins and Development Planning. In (J. Rose, Ed.), Environment Toxicology: Current Developments, Gordon and Breach Science Publishers, Amsterdam, pp. 81–107.

Siegel, F.R., Kravitz, J.H. and Galasso, J.J., 2000. Arsenic in arctic sediment cores: pathfinder to chemical weapons dump sites? Envir. Geol. 39: 705–706.

Siegel, F.R., Kravitz, J.H. and Galasso, J.J., 2001 a. Geochemistry of thirteen Voronin Trough cores, Kara Sea, European Arctic: Hg and As contamination at a 1965 timeline. Applied Geochemistry, 16: 19–34.

Siegel, F.R., Kravitz, J.H. and Galasso, J.J., 2001 b. Arsenic and mercury contamination in thirty-one 1965 cores from the St. Anna Trough, Kara Sea, Arctic Ocean. Envir. Geol., 40: 528–542.

Simpson, G.G. and Beck, W.S. (2nd edition), 1965. Life. Harcourt, Brace & Co., 869 p.

Sinclair, A.J., 1976. Probability Graphs in Mineral Exploration. Assoc. Explor. Geochemists, Rexdale, Ont., Canada, 75 pp.

Sjøbakk, T.E., Almli, B. and Steinnes, E., 1997. Heavy metal monitoring in contaminated river systems using Mayfly larvae: J. Geochem. Explor., 58: 203–207.

Skei, J.M., 1978. Serious mercury contamination of sediments in a Norwegian semi-enclosed bay. Marine. Poll. Bull., 9: 191–193.

Smith, 1999. Cadmium. In C.P. Marshall and R.W. Fairbridge (Eds.), Encyclopedia of Geochemistry, Kluwer Academic Publishers, Dordrecht, The Netherlands, pp. 50–51.

Snyder, 1999. Vanadium. In C.P. Marshall and R.W. Fairbridge (Eds.), Encyclopedia of Geochemistry, Kluwer Academic Publishers, Dordrecht, The Netherlands, p. 656.

Sondag, F., 1981. Selective extraction procedures applied to geochemical prospecting in an area of old mine workings. J. Geochem. Explor., 15: 645–652.

Sposito, G., 1983. The chemical forms of trace metals in soils. In (I. Thornton, Ed.), Applied Environmental Geochemistry, Academic Press, London, pp. 123–170.

Stanley, C.R., 1988. PROBPLOT – An interactive computer program to fit mixtures of normal (or log-normal) distributions using maximum likelihood optimization procedures. Spec. Vol. 14, Assn. of Exploration Geochemists, Rexdale, Ontario, Canada, 40 p., 1 diskette.

Stecko, J.R.P. and Bendell-Young, L.I., 2000. Contrasting the geochemistry of suspended particulate matter and deposited sediments within an estuary. Applied Geochem., 15: 753–775.

Stephenson, M., Turner, G., Pope, P., Knoght, A. and Tchobanoglous, G., 1980. The use and potential of aquatic species for wastewater treatment. Publ. No. 65, California State Water Resources Control Board, Sacramento, CA.

Strakhov, N.M., 1960. Principles of the Theory of Lithogenesis, Vols. 1 and 2. Moscow, Izd-vo Akad. Nauk SSSR, 240 p., (in Russian).

Swan, A.R.H. and Sandilands, M., 1995. Introduction to Geologic Data Analysis. Blackwell Science, Oxford, UK, 446 p.

Taylor, S.R. and McLennan, S.M., 1995. The geochemical evolution of the continental crust. Rev. Geophys. 33: 241–265.

Terry, N. and Zayed, A.M., 1994. Selenium volatilization in plants. In: (W.T. Frankenberger, Jr. and S. Benson, Eds.), Selenium in the Environment. Marcel Dekker, New York, pp. 343–367.

Tessier, A., Campbell, P.G.C. and Bisen, M., 1979. Sequential extraction procedure for the speciation of particulate trace metals. Anal. Chem., 51 (7): 844–851.

Thoming, J., Stichnothe, H., Mangold, S. and Calmano, W., 2000. Hydrometallurgical approaches to soil remediation - process optimization applying heavy metal speciation. Land Contam. and Reclam., 8: 19–31.

Thornton, I., 1993. Environmental geochemistry and health in the 1990s: a global perspective. Applied Geochem. Suppl. 2, 203–210.

Thornton, I., 1996. Sources and pathways of As in the geochemical environment: health implications. In (J.D. Appleton, R. Fuge and G.J.H. McCall, Eds.), Envir. Geochem. & Health, 113: 153-161 (Geol. Soc. Spec. Publ.).

Tsuchiya, K., 1978. Cadmium Studies in Japan – A Review. Kodansha Ltd., Tokyo, 376 p.

Turekian, K.K. and Wedepohl, K.H., 1961. Distribution of the elements in some major units of the Earth's crust. Geol. Soc. Amer. Bull., 72: 175–192.

Turk, J. and Thompson, G.R., 1995. Environmental Geoscience. Saunders College Publishing, Fort Worth, 428 p.

Urlings, L.G.C.M. 1990. In-situ cadmium removal-full scale remedial action of contaminated soil. Int'l Symp. on Hazardous Waste Treatment: Treatment of Contaminated Soils, Air & Waste Assn., and U.S. EPA Risk Education Laboratory, Cinncinatti, Ohio.

U.S. Bureau of Mines (1992). Manmade marshes: using nature to clean up polluted mine water. Wetlands, March 8, 3 pp.

U.S. Department of Agriculture Handbook 18, 1956 and September 1962 Supplement.

U.S. Environmental Protection Agency, 1992. Methods for the Determination of Metals in Environmental Samples. C.K. Smoley (Ed.), CRC Press, Boca Raton, FL, 339 p.

Van Ryssen, R., Baeyens, W., Alan, M. and Goeyens, L., 1998. The use of flux-corer experiments in the determination of heavy metal redistribuiton in and of potential leaching from the sediments. Water Sci & Tech., 37: 283–290.

Vangronsveld, J., Sterckx, J., Van Assche. and Clijsters, H., 1995. Rehabilitation studies on an old non-ferrous waste dumping ground: effects of revegetation and metal immobilization by beringite. J. Geochem. Explor., 52: 221–229.

Vardaki, C. and Kelepertis, A., 1999. Environmental impact of heavy metals (Fe, Ni, Cr, Co) in soils, waters and plants of Triada in Euboea from ultrabasic rocks and nickeliferous mineralization. Envir. Geochem. & Health, 21: 211–226.

Veleminsky, M., Laznicka, P. and Stary, P., 1990. Honeybees (*Apis mellifera*) as environmental monitors of heavy metals in Czechoslovakia. Acta Entomologica Bohemoslovaca, 87: 37–44.

Verner, J.F. and Ramsey, M.H., 1996. Heavy metal contamination of soils around a Pb-Zn smelter in Bukowno, Poland. In: (R. Fuge, M. Billet and O. Selinus, Eds.) Environmental Geochemistry, 3rd Int'l Symp., Krakow, Poland, Applied Geochem., 11: 11–16.

Veslind, P.A., Peirce, J.J. and Weiner, R.F., 1994. Environmental Engineering (3rd Ed.). Butterworth-Heinemann. Boston, MA, 544 p.

Vinogradov, A.P., 1962. Average contents of chemical elements in the principal types of igneous rocks of the Earth's crust. Geochemistry, pp. 641–664.

Vroblesky, D.A. and Yanosky, T.M., 1990. Use of tree-ring chemistry to document historical ground-water contamination events. Ground Water, 28: 677–684.

Wang, W.-X. and Fisher, N.S., 1999. Delineating metal accumulation pathways for marine invertebrates. Sci. Total Envir., 237/238: 459–472.

Waring, C. and Taylor, J., 1999. A new passive technique proposed for the prevention of acid drainage: GaRDS. In Mining Into The Next Century: Environmental Opportunities and Challenges. Proceedings 24th annual environmental workshop. Townsville, Minerals Council of Australia, pp. 527–530.

Warren, H.V., Delavault, R.E., Barakso, J., 1968. The arsenic content of Douglas Fir as a guide to some gold, silver, and base metal deposits. Can. Min. Metall. Bull., 61: 860–866.

Warren, H.V., Delavault, R.E., Peterson, G.R. and Fletcher, K., 1971. The copper and zinc content of trout livers as an aid in the search for favorable areas to prospect. In (R. Boyle, tech. ed.) Geochemical Exploration, Spec. Vol. No. 11, Canadian Institute of Mining and Metallurgy, pp. 444–450.

Watt, J., Thornton, I. and Cotter-Howells, J., 1993. Physical evidence suggesting the transfer of soil Pb into young children via hand-to-mouth activity. Applied Geochem., Suppl. 2, p. 269–272.

Watterson, A., 1998. Toxicology in the working environment. In (J. Rose, Ed.), Environmental Toxicology: Current Developments. Gordon and Breach Science Publishers, Amsterdam, pp. 225–252.

Webb, J.S., 1971. Regional geochemical reconnaissance in medical geography. In (H.L. Cannon and H.C. Hopps, Eds.), Environmental Geochemistry and Health. Geol. Soc. America Memoir 123, pp. 31–42.

Wilson, E.O., 1992. The Diversity of Life. Belknap Press of Harvard Univ. Press, Cambridge, 464 p.

Winder, L., Merrington, G. and Green, I., 1999. The tri-trophic transfer of Zn from the agricultural use of sewage sludge. Sci. Total Envir., 229: 73–81.

Wolverton, B.C., Barlow, R.M. and McDonald, R.C., 1976. Application of vascular aquatic plants for pollution removal, energy, and food production in a biological system. In (J. Tourbier and R.W. Pierson, Jr., Eds.), Biological Control of Water Pollution. Univ. Penn. Press, Philadelphia, PA, pp. 141–149.

Wong, J.S.H., Hicks, R.E. and Probstein, R.F., 1997. EDTA-enhanced electro-remediation of metal-contaminated soils. J. Hazardous Materials, 55: 61–79.

Yeargan, R., Maiti, I.B., Nielsen, M.T., Hunt, A.G. and Wagner, G.J., 1992. Tissue partitioning of cadmium in transgenic tobacco seedlings and field grown plants expressing the mouse metallothionein I gene. Transgenic Res., 1: 261–267.

Yin, Z., Yue, S and He, R., 1983. Characteristics of the environmental geochemistry [of selenium] and its relationship with "Keshan disease" in the plateau sand soil of Keshiketengqi, Inner Mongolia. Scientia Geographica Sinica, 3: 175–182 (in Chinese).

Younas, M., Khan, M.I., Ali, K., Shahzad, F. and Afzal, S., 1998. Assessment of Cd, Ni, Cu, and Pb pollution in Lahore, Pakistan. Environment Int'l, 24: 761–766.

Zhang, C., Selinus, O. and Kjellstrom, G., 1999. Discrimination between natural background and anthropogenic pollution in environmental geochemistry – Exemplified in an area of southeastern Sweden. Sci. Total Envir., 243/244: 129–140.

Zielhuis, R.L., 1979. General report: health effects of trace metals. In (E. DiFerrante, Ed.), Trace Metals: Exposure and Health Effects. CEC and Pergamon, pp. 239-247.

Epilogue

How green is my environment? The greener, denser and more widespread it is the better off are we on land and in water. Healthy vegetation means healthy foul, fish and animal populations and these in turn convey health to human populations. Certainly there are diseases which afflict all populations and we cannot escape those of bacterial or viral nature. Nor may we be able escape those that are genetically inherited. That is until the recently completed mapping of the genome leads to research through which genes that can stimulate disease can be modified.

However, we can attend to our environment by cleaning up the toxins that have originated through anthropogenic activities and that have had ready access to the living environment. This has begun at an increasing number of sites worldwide, but the proverbial global surface has been only slightly scratched. The amount of investment and interest in cleanup by past industrial polluters and governmental agencies is growing. In some cases this is driven by NGO and political pressures and with an eye towards long-term goodwill and ultimately profits. In other cases, it is because industries can benefit. For example, some mining companies have been reworking waste spoils from old gold mines with profit motives while at the same time reinvesting part of the profits in remediating acid mine drainage and heavy metals contamination problems caused by the waste tailings and abandoned mines.

We can work towards eliminating the introduction of toxins originating from contemporary human activities to the living environment. Among these toxins that can poison the foodweb and consumer populations are the metals we have discussed in this text. Good progress is being made globally with the introduction of legislation and its enforcement resulting in retrofitting of many industries. In addition, the development of new industries is designed to greatly limit the release of heavy metals to the environment. One result of both approaches has been a great reduction of contaminant atmospheric emissions worldwide.

The future is promising. More research into remediation technologies is supported well by governmental and industrial contracts. More and better technologies are being developed that will increase the efficiency of metal toxin cleanup and decrease the time necessary for remediation to natural environmental conditions. Working against the positive movement is the growth of human populations with their needs for natural resources to satisfy normal aspirations and growing acquisitive power. A balance between growth and technologies that keep environments free of impacting chemicals is achievable but only with responsible investment. Ultimately, sustainability may be achieved so that clean, vital ecosystems pass on from one generation to the next while supplying the needs of populations and at the same time bettering the quality of life globally.

Subject Index